Locke and Cartesian Philosophy

Locke and Cartesian Philosophy

EDITED BY
Philippe Hamou
and Martine Pécharman

OXFORD
UNIVERSITY PRESS

OXFORD
UNIVERSITY PRESS

Great Clarendon Street, Oxford, OX2 6DP,
United Kingdom

Oxford University Press is a department of the University of Oxford.
It furthers the University's objective of excellence in research, scholarship,
and education by publishing worldwide. Oxford is a registered trade mark of
Oxford University Press in the UK and in certain other countries

First Edition published in 2018
Impression: 1

Published in the United States of America by Oxford University Press
198 Madison Avenue, New York, NY 10016, United States of America

British Library Cataloguing in Publication Data
Data available

Library of Congress Control Number: 2017958744

ISBN 978–0–19–881503–7

Printed and bound by
CPI Group (UK) Ltd, Croydon, CR0 4YY

Contents

Abbreviations

A	German Academy of Sciences, ed. *G. W. Leibniz: Sämtliche Schriften und Briefe*. Darmstadt and Berlin: Berlin Academy, 1923–.
AG	*An Early Draft of Locke's Essay: Together with Excerpts from his Journals*, ed. Richard I. Aaron and Jocelyn Gibb, Oxford: Clarendon Press, 1936.
AT	C. Adam and P. Tannery, eds., *Oeuvres de Descartes*, 12 vols. Paris: Vrin, 1897–1913; repr. Paris, Vrin, 1964–76.
BL	British Library.
CSM	J. Cottingham, R. Stoothoff, and D. Murdoch, trans., *The Philosophical Writings of Descartes. Vol. I & II*. Cambridge: Cambridge University Press, 1985.
CSMK	J. Cottingham, R. Stoothoff, D. Murdoch, and A. Kenny, trans., *The Philosophical Writings of Descartes. Vol III. The Correspondence*. Cambridge: Cambridge University Press, 1991.
DM	Nicolas Malebranche, *Dialogues on Metaphysics and on Religion.*
Draft A, Draft B	John Locke, *Drafts for the Essay Concerning Human Understanding and Other Philosophical Writings*, vol. 1, Drafts A & B ed. Peter H. Nidditch and G. A. J. Rogers. Oxford: Clarendon Press, 1990.
E	John Locke, *An Essay Concerning Human Understanding*, ed. P. H. Nidditch. Oxford: Clarendon Press, 1975.
EEH	John Locke, *Essai concernant l'entendement humain*, trans. Pierre Coste, ed. E. Naert. Paris: Vrin, 1974.
JS	N. Jolley, ed., and D. Scott, trans., *Nicolas Malebranche: Dialogues on Metaphyics and on Religion*. Cambridge: Cambridge University Press, 1997.
LO	T. M. Lennon and P. J. Olscamp, eds., *Nicolas Malebranche: The Search After Truth*, rev. edn. Cambridge: Cambridge University Press, 1997.
LC	E. S. De Beer, ed., *The Correspondence of John Locke*, 8 vols. Oxford: Clarendon Press, 1976–89.
LL	John Harrison and Peter Laslett, *The Library of John Locke.* Oxford: Clarendon Press, 1971.

LW	John Locke, *The Works of John Locke*, 10 vols. London: Thomas Tegg, 1823, repr. Aalen, 1963.
NE	G. W. Leibniz, *New Essays on Human Understanding.*
OCM	A. Robinet, ed., *Œuvres complètes de Malebranche*, 20 vols. Paris: Vrin, 1958–67.
P	René Descartes, *Principles of Philosophy.*
RB	P. Remnant and J. Bennett, trans., *G.W. Leibniz: New Essays on Human Understanding*, rev. edn. Cambridge: Cambridge University Press, 1996.
Reasonableness	John Locke, *The reasonableness of Christianity as delivered in the scriptures*. Edited with an introduction, notes, critical apparatus and transcriptions of related manuscripts by John C. Higgins-Biddle. Oxford: Clarendon Press, 1999. (The Clarendon edition of the works of John Locke.)
SAT	Nicolas Malebranche, *The Search After Truth.*
T	David Hume, *A Treatise of Human Nature.*

Contributors

PETER R. ANSTEY is Professor of Philosophy at the University of Sydney. He specializes in early modern philosophy with a particular focus on the philosophy of John Locke, experimental philosophy, and the philosophy of principles. He is the author of *John Locke and Natural Philosophy* (Oxford, 2011) and editor (with Lawrence Principe) of the forthcoming Clarendon edition of Locke's writings on natural philosophy and medicine.

ANDREAS BLANK is Visiting Associate Professor in philosophy at Bard College Berlin. His publications include *Der logische Aufbau von Leibniz' Metaphyik* (De Gruyter, 2001), *Leibniz: Metaphilosophy and Metaphysics, 1666–1686* (Philosophia, 2005), *Biomedical Ontology and the Metaphysics of Composite Substances, 1540–1670* (Philosophia, 2010) and *Ontological Dependence and the Metaphysics of Composite Substances, 1540–1716* (Philosophia, 2015). He is currently completing a book on presumptions and early modern practical rationality.

MARTHA BRANDT BOLTON, Professor of Philosophy at Rutgers University, has written on many topics in the history of early modern metaphysics and theory of cognition including Descartes on thinking as a principle attribute, Locke on the sorts of things which remain the same in time, and the engagement of Locke's *Essay* by Leibniz's *New Essays*.

LISA DOWNING is Professor of Philosophy at the Ohio State University. She has published widely in early modern philosophy (on Descartes, Malebranche, Boyle, Locke, Berkeley, and Newtonianism), especially on connections among physics, metaphysics, and philosophy of science in the period.

PHILIPPE HAMOU is Professor of Philosophy at Université Paris-Nanterre. He has published on early modern philosophy and science, with special focus on Galileo, Locke, Newton, vision, and visuality. He is currently completing a book on Locke's concept of mind.

MATTHIEU HAUMESSER teaches philosophy at a public high school in Cergy-Pontoise and at the Paris Institute of Political Studies. He works on modern philosophy, especially on Locke and Kant. His publications include *Kant: De L'Amphibologie des Concepts de la Réflexion* (Paris: Vrin, 2010), in which he studies Locke's influence on Kant's critical philosophy.

JAMES HILL is a privatdozent at Charles University in Prague and a fellow of the Philosophy Institute in the Czech Academy of Science. He has published widely on early modern philosophy, including *Descartes and the Doubting Mind* (Bloomsbury, 2012). He is currently working on a monograph on George Berkeley.

LAURENT JAFFRO is Professor of Moral Philosophy at Panthéon-Sorbonne University, Paris. His work focuses on moral theory and the history of early modern British philosophy. He has published in particular on the third Earl of Shaftesbury, John Toland, George Berkeley, and Thomas Reid.

NICHOLAS JOLLEY is Emeritus Professor and Research Professor of Philosophy at the University of California, Irvine. His publications include *Locke: His Philosophical Thought* (OUP, 1999) and *Locke's Touchy Subjects: Materialism and Immortality* (OUP, 2015). His most recent book is *Toleration and Understanding in Locke* (OUP, 2016).

DENIS KAMBOUCHNER is Professor of the History of Early Modern Philosophy at the Université Paris 1 Panthéon-Sorbonne. He has written numerous studies on Descartes (recently published: *Descartes n'a pas dit*, Paris, Les Belles-Lettres, 2015) and is now the chief editor of the new edition of Descartes's Complete Works (Gallimard, in progress). He is currently completing a comprehensive study of the *Metaphysical Meditations*.

J. R. MILTON is Professor Emeritus of the History of Philosophy at King's College London, and General Editor of the Clarendon Edition of the Works of John Locke. He has published widely on Locke and other topics in early modern philosophy, and is currently finishing work on two volumes for the Clarendon Edition: *Literary and Historical Writings* and *Drafts of the Essay concerning Human Understanding and other Philosophical Writings*, volume 2.

MARTINE PÉCHARMAN is Director of Research at the Centre National de la Recherche Scientifique (CNRS, Paris). She has written broadly on early modern logic (with a special focus on Hobbes and on Port-Royal), metaphysics, and ethics. She has also produced critical editions of Hobbes, Bayle, Condillac.

CATHERINE WILSON is Anniversary Professor of Philosophy at the University of York. She has published widely in early modern philosophy and is working on a book on Kant and the life and human sciences.

Introduction

Philippe Hamou and Martine Pécharman

Locke's relationship to Descartes and Cartesian philosophy has long been an important theme in Lockean studies.[1] Until quite recently however, the historiography has suffered from an almost exclusive focalization on epistemological issues. The chapter on Descartes in James Gibson's *Locke's Theory of Knowledge* is a good illustration of how Locke's connection to Descartes has been usually interpreted during the past century. According to Gibson, who compares a striking passage from the *Regulae*[2] to Locke's own general statements of intent in the *Essay*, Locke's and Descartes's philosophical aims were essentially similar: they both set out to enquire into the sources of knowledge, in order to determine what can be known with certainty. Their proposed *methods*, drawing on the *consciousness* that we have of our own *ideas*, and on the *intuitive* perception of the relation between them, also present striking resemblances. Gibson considered, however, that Locke went further, and on more secure grounds, than Descartes himself, being more rigorous in his treatment of the epistemological problem (which means, for Gibson, more careful to avoid metaphysical conundrums),

[1] General contributions on Locke's relation to Descartes include Frank Thilly, 'Locke's Relation to Descartes' (1900); James Gibson, *Locke's Theory of Knowledge and its Historical Relations* (1917); Peter A. Schouls, 'The Cartesian Method of Locke's *Essay*' (1975); Thomas M. Lennon, *The Battle of the Gods and Giants* (1994); G. A. John Rogers, 'Descartes and the Mind of Locke' (1998); Paul A. Schuurman, *Ideas, Mental Faculties and Method* (2004) and 'Descartes, René' (2010), Schmaltz (ed.), *Receptions of Descartes* (2005); Nicholas Jolley, 'Lockean Abstractionism versus Cartesian Nativism' (2008), and 'Dull Souls and Beasts' (2010); Lisa Downing, 'Locke and Descartes' (2015).

[2] In the *Regulae*—published only in 1701—the passage which, according to Gibson, 'seems almost verbally to anticipate' Locke's critical questioning of the bounds of knowledge reads: 'Now there does not arise here any problem the solution of which is of greater importance than that of determining the nature of human knowledge and how far it extends; two points which we combine into one and the same enquiry, which it is necessary first of all to consider in accordance with the rules given above. This is a question which one must face once in one's life, if one has ever so slight a love of truth, since it embraces the whole of method, and as it were the true instruments of knowledge. Nothing seems to me to be more absurd than to discuss with boldness the mysteries of nature, the influence of the stars, and the secrets of the future, without having once asked whether the human mind is competent to such enquiries' (*Regulae ad directionem ingenii*, VIII, quoted in Gibson, *Locke's Theory of Knowledge*, pp. 207–8).

and pointing out where precisely Descartes had gone wrong. As Gibson wrote in the conclusion of his chapter on Locke and Descartes:

> In the attempt to determine fundamental questions of fact in an *a priori* manner, apart from any reference to experience, and in the tendency to offer an exposition of conceptions in place of a synthetic demonstration, there was evidence that after all Descartes had not completely emancipated himself from the toils of the scholastic logic. And since these features were precisely those which a theory of innateness was designed to support, the defects of method and the presence of the offending theory could hardly fail to be connected in Locke's mind.[3]

Gibson had earlier made an interesting statement about Descartes's 'influence' on Locke:[4]

> Without the influence of the Cartesian view of knowledge and the Cartesian conception of self-consciousness, it is not too much to say that the *Essay*, as we know it, would never have been written. At the same time, we shall find that the way in which Locke develops the view of knowledge which he found in Descartes, and the very different use to which he puts the conception of self-consciousness, suffice to negative at once the suggestion of any want of originality in his fundamental positions. So freely indeed, does he transform the Cartesian principles that the existence of any positive relation of dependence upon them has frequently been ignored by the historian of philosophy, and the positions of Descartes and Locke have been set in antithetical opposition to each other.

Gibson is certainly right when he criticizes the 'antithetical' view of the relationship that prevailed among historians and philosophers of the eighteenth and nineteenth century, and most contributors to this volume would concur with the view that a number of Cartesian themes are, indeed, freely and subtly appropriated by Locke. A more debatable point in Gibson's approach is the suggestion that Locke and Descartes were basically pursuing the same end, seeking to provide safe grounds for scientific knowledge, using the same method of certainty through the 'way of ideas'. Gibson seems to consider as positive a view that was put forward with negative intent by John Sergeant, Edward Stillingfleet, or Henry Lee as their main argument against Locke's 'Cartesian' doctrine of knowledge. In so doing, he expresses what will become an influential historiographical conception, namely that Descartes and Locke form together the spearhead in the 'epistemological turn' of early modern philosophy.[5] Since then indeed, the theory of knowledge has been the main locus for the comparison of the two authors.[6]

[3] Gibson, *Locke's Theory of Knowledge*, p. 232. [4] Ibid. p. 207.

[5] On the so-called 'epistemological turn' see Rorty, *Philosophy and the Mirror of Nature*; and for a critical account of Rorty's historiography, Hatfield, 'Epistemology and Science in the Image of Modern Philosophy: Rorty on Descartes and Locke'.

[6] See in particular Schuurman, *Ideas, Mental Faculties, and Method*. The idea that Locke basically appropriated the whole of Descartes's 'method' and, separating it from metaphysics, made it palatable for empirical minded eighteenth-century philosophers, is developed at length in P. A. Schouls's publications. See, *inter alii*, his 1975 paper 'The Cartesian Method of Locke's *Essay concerning Human Understanding*', and the book that followed: *The Imposition of Method: A Study of Descartes and Locke*.

In bringing together the several contributions of this volume, we would like to advocate for a shift of emphasis. As chapters in this volume amply show, there is much to learn from the comparison of Locke's and Descartes's positions on physical, metaphysical, and religious matters. Their conflicting claims on issues such as cosmic organization, the qualities and nature of bodies, the nature of ideas, the substance of the soul ... are relevant, not only in their own right, to take the full measure of Locke's complex relation to Descartes, but also as they allow a better understanding of the epistemological debate.

Let us consider these various points somewhat more closely, starting with the old antithetical view. There is of course, something inescapable in it. In many respects Locke and Descartes were very different men, besides being obviously very different philosophers. Their religious breeding and convictions (if not their 'essential religiosity'[7]), their attitudes towards political involvement, their prose and style were altogether distinct, almost opposite. Both spent a long and fruitful period of exile in Holland, but for Descartes it was by personal choice, whereas Locke was fleeing a threat of prison in the heated political climate that followed the discovery of the Rye House plot. On many doctrinal philosophical points, Locke held specifically anti-Cartesian theses. He did not think that the soul always thinks, that we have innate ideas, that we have a positive idea of infinity,[8] nor that we could have an idea of a *chiliaëdron*[9] that is not an image; he denied that our idea of body and our idea of extension are one and the same, and so on. On even more numerous other points, Locke expressed serious doubts about well-known Cartesian doctrines. He was reluctant to give much weight to the ontological proof of God's existence.[10] He had trouble with the certainty of the so-called 'dualistic' account of matter and spirit, even suggesting in IV. iii. 6 that, because our limited knowledge is unable to master the puzzles raised by our ideas of matter and thought, the soul might therefore be material. He remained entirely unconvinced by the doctrine of beast-machines,[11] etc. Voltaire's *Letter on Mr Locke* was perhaps the earliest and most influential expression of the antithesis between, on the one hand, Moderns such as Descartes and Malebranche who still belonged to the 'multitude of reasoners' writing 'the romance of the soul' and, on the other hand, the wise and modest Locke who, as an 'excellent anatomist', was the first to write its 'history'.[12] Since then it has been common practice to represent the two authors as personifications of some of the major

[7] See in this volume Catherine Wilson's contribution.

[8] See E II. xvii.

[9] See Descartes, *Meditatio Sexta*: AT 7, pp. 72–3 and Locke, E II. xxix.14.

[10] See E IV. x. 7. This point was developed in Locke's manuscript 'Deus Des Cartes's proof of a god from the Idea of necessary existence examined 1696' (MS Locke c. 28, fos. 119^r–120^v).

[11] See E II. xi. 11 ('if [Brutes] have any *Ideas* at all, and are not bare Machins (as some would have them) we cannot deny them to have some Reason') and II. i. 19 ('they must needs have a penetrating sight, who can certainly see, that I think, when I cannot perceive it my self, and when I declare, that I do not; and yet can see, that Dogs or Elephants do not think, when they give all the demonstration of it imaginable, except only telling us, that they do so').

[12] Voltaire, *Letters Concerning the English Nation*, Letter XIII, pp. 98–9.

antinomies of early modern philosophy, the enduring battle of Gods and Giants,[13] rationalism and empiricism, nativism and empiricism,[14] free will and determinism. Locke himself seems to have been partly responsible for the invention of the antagonistic view, and the spreading in England of a rather common caricature of Descartes. Of all the authors referred to directly or (more often) indirectly in the *Essay*, Descartes is certainly the most conspicuously present. But his name (or the word *Cartesian(s)*), which recurs significantly often in a work where very few proper names are mentioned, is almost never mentioned in praise.[15] As Pierre Coste sourly remarked in the footnotes to his French translation of the *Essay*, whenever Descartes comes to the fore, Locke's judgement, usually sound and measured, appears somewhat twisted, often verging on caricature. In his footnotes to chapter II. i for instance, Coste was dissatisfied with Locke's way of presenting the Cartesian thesis that the soul thinks always.[16] Also, in his footnotes to II. xiii, he stressed Locke's rather unfair attribution to Cartesians of the thesis that sensible qualities are all inseparable from extension, in order to argue for the distinction of the ideas of space and body.[17] In some cases the Lockean arguments are simply irrelevant, attacking doctrines that are not really Cartesian, and in others, they tend to distort or harden the Cartesian position, so that it can easily be identified with its most extreme and abhorrent consequences—as for

[13] For instance, in his book *The Battle of the Gods and Giants*, Thomas M. Lennon contends that Locke supports the cause of Gassendi versus Descartes, renewing the perennial philosophical battle characterized in Plato's *Sophist* as the struggle of materialism (Giants) against idealism (Gods). In Lennon's opinion, the whole of Locke's *Essay* constitutes an 'anti-Cartesian polemic', notably, in the light of the two-persons argument in II. i. 12 against the Cartesian thesis that the soul always thinks, the 'long treatment of personal identity' added to the second edition of the *Essay* in II. xxvii cannot be viewed 'just as a topic of independent philosophical importance' (p. 168).

[14] The question whether Descartes was or was not the main target of the first book of the *Essay*, on Innate ideas and Principles has been scrutinized in many publications. See, for example, Thilly, ' Locke's Relation to Descartes', Yolton, *Locke and the Way of Ideas*.

[15] A rare exception occurs in E. III. iv. 10, where the Cartesian distinction between the cause ('*Des Cartes*'s Globules') of our sensation of light and this sensation itself supports Locke's statement that the names of simple ideas cannot be defined.

[16] See Coste's footnotes to E II. i. 16, EEH p. 69 ('I do not think that those against whom Mr Locke is fighting here have ever considered maintaining that the soul of a man is more separated from the body when the man sleeps than when he is awake' ['Je ne pense pas que ceux que Mr. Locke combat ici, se soient jamais avisés de soutenir, que l'ame de l'Homme soit plus séparée du corps pendant que l'Homme dort, que pendant qu'il veille'] and to II. i. 17 (against the attribution to Cartesians of the thesis of the pre-existence of the soul). Coste actually defended a form of 'Pyrrhonism' against both Descartes and Locke ('un peu de Pyrrhonisme ne siérait point mal, à mon avis', p. 69).

[17] See Coste's footnote to E II. xiii. 25, and to E II. viii. 14: 'Mr. Locke…seems to have entirely forgot how Cartesians explain Sensible Qualities…It is difficult to understand what drove Locke to spin out this long argument against the Cartesians. He obviously has something against them here, talking as he does of the ideas of tastes or odors as if the Cartesians believed that they were inherent qualities of bodies. But it is very certain that, long before Locke even thought of composing his book, the Cartesians had demonstrated that the ideas of tastes and odors are solely in the minds of whoever tastes the bodies that are called tasty or smells the ones that are called odoriferous, and that so far from including in them any idea of extension, these ideas are excited in our souls by something in bodies that is entirely unrelated to them. When I came to translate this passage of the Essay, I saw Mr. Locke's mistake, and I warned him about it—but I could not obtain from him, that the opinion he attributed to the Cartesians was directly opposed to the one they held and most evidently proved, and which he himself endorsed in his book' [our translation].

example, when Locke identifies the Cartesian dualistic account of mind and body with a quasi-Platonistic account of soul–body dissociation, drawing on strange thought-experiments of soul transmigrations between animal–machine bodies, in order to show the seemingly appalling consequences of Descartes's doctrine of pure thought.[18] At a time when Descartes's persona and Descartes's thought still had a very strong hold on European minds, it can appear as if Locke (who was surely not immune to a certain nationalistic prejudice against the 'French Philosophers'[19]) was taking it upon himself to provide the tools for dismantling the Cartesian statue, and French philosophical pre-eminence.

However, even though Locke himself appears willing to plead for it, the common antithetical reading of the relationship between the two authors must be taken with some caution. Nationalistic prejudices and caricatures aside, Locke's relationship to Cartesian philosophy seems to be far more complex. As suggested earlier in Gibson's quotation, Locke freely incorporated into his philosophical insight, making them almost organic constituents, a number of Cartesian themes, concepts, and methodological commitments. For example, Locke's way of philosophizing certainly perpetuates a style and an ideal that owe much to Descartes. Locke never made a mystery of Descartes's role in awakening his interest in philosophy. As he once told Lady Masham, Descartes's writings were the first to give him 'a relish of Philosophical Studys'.[20] Similarly, in his controversy with Stillingfleet, he acknowledged a 'great obligation' to Descartes in that he owed him his 'first deliverance from the unintelligible way of talking of the philosophy in use in the Schools'.[21] Locke, perhaps more than any other in the century, developed and transmitted Descartes's legacy of a philosophy written in plain language, addressed to the common reader, and deliberately avoiding the Scholastic jargon. More importantly, Locke adopted Descartes's decision to treat philosophical questions in a 'first-person' perspective. Admittedly, the use of the first person (singular or plural) is certainly less systematic in the *Essay* than it is in the *Meditations* or the *Discourse on Method*. But nevertheless, the same philosophical idea is here at the very heart of both enterprises: truth is always firstly encountered as subjective certainty. All philosophical questions are to be treated, not through

[18] See E II. i and II. xxvii.

[19] See the testimony of Pierre Coste, as reported by his friend and biographer, De La Motte: 'Besides M. Locke was growing more and more reserved towards M. Coste who did not show him enough consideration. For example, Locke never missed the least opportunity to run down Descartes, father Malebranche, the art of thinking and the most famous Parisian academicians. M. Coste defended them without hesitation, while always keeping to the strictest rules of courtesy, defending only what he thought defensible. There are a thousand insinuations in Locke's Essay on the Understanding against the Cartesians, some of which M. Coste has picked out in the annotations of his last editions of the Essay. Some people found it was the wrong thing to do, whereas most of his friends urged him to make more of such annotations', translated from Charles de la Motte, *La Vie de Coste et anecdotes sur ses ouvrages*, in Locke, *Que la religion chrétienne est très-raisonnable*..., pp. 231–60.

[20] Lady Masham to Jean Le Clerc on 12 January 1705 (quoted by Schuurman in *Ideas, Mental Faculties, and Method*, p. 16).

[21] LW 4, p. 48.

dialectical considerations of the best available opinions, nor through deductions from general principles or maxims, but as they appear to a singular subjective experience, to an unprejudiced mind, dealing with how things *look* to itself, how they appear to be—or what they are *in its ideas*. No doubt, the *Essay* remains a very different book from the *Meditations*. Locke has no patience for universal doubt, and does not think highly of its epistemic virtues. Rather than a strict demonstration, in which nothing is admitted if not analytically deduced from first truths, Locke's reader is invited to follow the sinuous and somewhat rambling discourse of the *Essay*, interspersed with digressions and tacit suggestions of more or less probable opinions—a discourse that Locke presents as the very image of his own wandering and curious mind, open to whatever comes into view—and sometimes even surprised by his own discoveries.[22] This no doubt reveals crucial differences—but once again, Locke and Descartes share the same ground, the same conviction that philosophy starts in some sense with the history of one's own mind, and that we, as philosophers, have the epistemic duty to build on a ground that is *all ours*, a duty to be the sole authors and warrants of whatever we hold to be true.[23]

It must be emphasized again that many theoretical options that Locke followed in the *Essay* would not have been *possible* without Descartes. This of course is quite obvious on issues related to mind and knowledge, where Locke uses such unmistakable Cartesian concepts as 'clear and distinct ideas',[24] 'consciousness', 'thinking things', 'intuitive' and 'demonstrative' knowledge, and sometimes characterizes them in terms that are literally taken from Descartes. But even on topics where Locke is overtly attacking Descartes, for example on substance and mode, on space and extension, on the freedom of the will, it can be shown (see e.g. the chapters by Bolton, Hill, and Kambouchner) that Locke's positions depend at a constitutive level on Cartesian premises, and could not have been produced without them. For example, his criticism of innate ideas is grounded in a strict Cartesian definition of thought as conscious thinking. His anti-Cartesian definition of bodies reposes on a broadly Cartesian view of material substance as a fully actualized and undifferentiated stuff, of which all bodies are constituted. On some occasions, it seems that Locke is exploring theoretical possibilities that Descartes himself had opened up but had not really wanted to investigate. For example, a case could be made that Locke's view of *ideas* is a direct descendant of the theory of sensation that can be found in the *Treatise on Man*, where it is said that *ideas* are kinds of images or pictures that are projected into the brain and there become

[22] See E The Epistle to the Reader, pp. 7–8: 'when I first put Pen to Paper, I thought all I should have to say on this Matter, would have been contained in one sheet of paper; but the farther I went, the larger Prospect I had: New Discoveries led me still on, and so it grew insensibly to the bulk it know appears in.'

[23] For a characterization of the Lockean 'epistemological individualism' and its connection with the denial of innate ideas, see Jolley, *Locke: His Philosophical Thought*. Jolley writes: 'Locke's commitment to epistemological individualism . . . is something he shares with Descartes, but in one way he goes beyond his great predecessor. Or to put the point another way, he seeks to be more faithful to this commitment than Descartes himself' (p. 171).

[24] The expression was perhaps too conspicuously Cartesian. Locke decided to replace it by *determinate* or *determined* ideas in the fourth edition of the *Essay* (E The Epistle to the Reader, p. 13). See Pécharman, 'Le Problème de la distinction des idées', in Hamou and Gaudemar, eds., *Locke et Leibniz*, 13–45.

the object of sensory awareness. Of course, to Descartes, *ideas* in this sense are not what *ideas* in the intellectual sense are. In later texts, to avoid equivocation, Descartes tended to drop the language of *ideas* when talking about sensory images. But for anyone who strongly doubts (as Locke did) that something like 'pure intellectual thoughts' can exist, *Cartesian sensations* are indeed the only immediate content of the mind that we are left with. The challenge for a Lockean epistemology would therefore be to make sense of true knowledge, even mathematical knowledge, with these Cartesian sensory images and with them alone.

So on these issues and on a number of other topics, Locke drew on Descartes's concepts and doctrines with a kind of casual selectiveness, borrowing from them, as from a tool box, what he needed, with little or no consideration of their original purposes, and sometimes of course in direct opposition to them. This kind of selective or idiosyncratic appropriation is perhaps typical of how philosophers work and capitalize on others. But in the case of Locke and Descartes, the ties are so strong that there is a sense in which it can be said that the *Essay* belongs to the history of 'Cartesianism'. Naturally, the term should not be understood as describing a 'school', identified by a set of fixed and intangible theses—such a school never actually existed. The term should be taken, rather, as a general label for the various ways Descartes's philosophy was used, selectively interpreted and transformed in the course of the long seventeenth century.

This in turns explains the need to consider Locke's complex appropriation of Descartes in a larger context, involving other subject matters, and other actors and perpetuators of Cartesianism, such as the Port-Royal authors, Malebranche, Clauberg, etc. Not only did these authors contribute to the diffusion of Cartesian ideas in Europe, and notably in England, they also put these ideas into use, applying them to questions and fields that were not directly addressed by Descartes—specifically, in the case of Port-Royal and Clauberg, to linguistics and logic. As his journal of travels in France shows, Locke read many of these authors, perhaps as extensively as he read Descartes. Locke's journal in Paris on 7 March 1678 reproduced a short anonymous writing entitled *Methode pour bien etudier la doctrine de Mr de Cartes*, which recommended reading Cartesians, not only Descartes: 'apres avoir bien conceu la maniere de philosopher dans sa methode on peut lire sur le sujet de la Logique celle que nous ont donnee Mrs de Port Royal qui est un ouvrage le plus accompli qui ait encore paru en ce genre et faire l'application des 4 regles de Des Cartes sur les quatre parties qu'elle contienne'. The unknown author of the *Methode* added: 'On peut encore lire la dessus *la Logique de Clauberge* qui a servi comme de fondement a celle-la et un autre traité sur le méme sujet que M^r^ *Du Hamel* a intitulé *de Mente Humana*.'[25] On several topics—especially

[25] See AG p. 107. We would like to thank J. R. Milton for drawing our attention to the unlikelihood that Locke, who was not fluent in French, wrote this *Methode*. According to Bonno (*Les Relations intellectuelles de Locke avec la France*, p. 91), Pierre-Sylvain Régis might be the author of the bibliographical references contained in this document. Lough, in his *Locke's Travels in France*, p. 60, quotes an entry in another Locke's notebook in 1678 ('Regis. proch la Greve a la rue de Tissandery a la test d'or St. Augustin place Daulphin'), which he understands as meaning that Locke had then contacts in Paris with Pierre-Sylvain

on words, on propositional attitudes (such as assent or negation), on the status of maxims and principles—the Cartesians were more direct interlocutors for Locke than Descartes was. Moreover, of the French philosophers, it was probably Malebranche rather than Descartes who came to the forefront of Locke's philosophical and polemical interests at the end of his life.[26] The inclusion in this book of three studies dealing with these so-called 'Cartesians' will certainly help broaden our understanding of Locke's uses and criticisms of Cartesian ideas.

We suggested earlier that epistemology (the theory of knowledge) has for a long time been the main focus of studies dealing with Locke's relationship to Descartes. Contemporary concerns probably buttressed the interest of interpreters from the last century in these sorts of questions. At a time when Locke was held as one of the respectable ancestors of the modern 'logical-positivist' theory of knowledge, it might have seemed especially interesting, and perhaps somewhat perplexing, to consider how he could have managed to combine a Cartesian-like 'foundational' intent with a strong commitment to 'empiricism'. It seems that we know better now: Locke's commitment to the idea that all knowledge is founded on experience is not foundationalist, if by this we mean that it may serve to unify the sciences or to make apparent their abstract logical structure. In the last chapter of the *Essay*, Locke clearly distinguishes between the intellectual provinces, showing that it is one thing to study the instruments (or signs) we use to know things (ideas and words), and another, altogether different, to study the things themselves 'as they are in their proper beings'.[27] If the 'Under-Labourer' (the Lockean philosopher) is useful to the 'Master-Builders' (the natural philosophers),[28] it is in clarifying what they are up to, and not in offering to them whatever principles or hard, incontrovertible data they are supposed to need to construct their edifice. In this, Locke also appears quite far removed from the Descartes of the *Meditations* and *Principles*, if not from the Descartes of the *Regulae*.

Besides, the very idea that Locke's philosophy is promoting an 'empiricist' view of knowledge against the 'rationalist' view of the Cartesians raises perhaps as many problems as it supposedly solves. 'Empiricism'/'rationalism' are not categories of

Régis. Our suggestion—confirmed by Antonella Del Prete—would be rather that this 'Regis' was the Calvinist Pierre Régis (1656 Montpellier–1726 Amsterdam), who came to Paris in 1678 just after gaining his medical degree in Montpellier, to attend lessons taught by the anatomist Joseph Duverney and the chemist Nicolas Lémery. Fontenelle's *Eloge de M. Régis* indicates that Pierre-Sylvain Régis stayed in Montpellier until 1680. So, the identity of the author of the *Methode pour bien etudier la doctrine de Mr de Cartes* remains a mystery.

[26] The manuscript of A*n Examination of P. Malebranche's Opinion of seeing all Things in God* was partially published in 1706 by Peter King in Locke's *Posthumous Works*. Locke's *Remarks upon Some of Mr Norris's Books, wherein he asserts P. Malebranche's opinion of our seeing all Things in God,* dealing with Norris's *Cursory Reflections on the Essay* (1690[1], 1692[2]) and with his *Reason and Religion* (1689[1], 1693[2]), was published in 1720 by Pierre Des Maizeaux.

[27] Cf. IV. xxi.

[28] E The Epistle to the Reader, pp. 9–10. In the years before the publication of the *Essay*, Locke reviewed for the *Bibliothèque universelle et historique* major books by two of these 'Master-Builders': Robert Boyle's *De specificorum remediorum cum corpusculari philosophia concordia* (May–August 1686) and Isaac Newton's *Philosophiae naturalis Principia mathematica* (January–March 1688).

Locke's time; their application to seventeenth-century thinkers was the result of retrospective and often polemical readings, through Kantian and post-Kantian glasses. Although Locke certainly says that sensory experience provides the *material* of all knowledge—yet not alone, since the mind's reflection on its operations constitutes a second mode of '*Experience*'[29]—his own definition of knowledge as the act through which the mind perceives the agreement of ideas appears to be a rather intellectualist one (if the term is any improvement); and Locke's commitment to reason (that is, to demonstration and proofs) was certainly no weaker than Descartes's.

Another reason why commentary on Locke has long focused almost exclusively on epistemological questions is Locke's own restraint regarding physical and metaphysical issues. Locke made quite clear from the very start of the *Essay* that, concerning the mind, he did not want to 'meddle' with any 'physical consideration'.[30] Correlatively, in the last chapter of the *Essay*, he explained that in his division of sciences, the term *physics* or 'natural philosophy' is to be understood in an 'enlarged Sense',[31] including whatever concerns the *proper beings* of things, either corporeal or spiritual. Among such physical and/or metaphysical questions are the true constitution of matter, the explanation of its various powers and activities, its cosmic arrangement and motions, the existence of void space, the relation between mind and bodies, the essence and mechanisms of the mind, the nature of ideas and how much they depend on matter, etc. All these questions were obviously central to Descartes, and Descartes's stance on them, his particular brand of mechanism, his identification of body and extension, his vortex theory, his concept of the mind as an immaterial substance, etc. were of course hotly debated in the seventeenth century. As Locke claimed to remain agnostic on physical matters, he might appear not to have wanted to side either with or against Descartes on these issues. But this is certainly false: on all these questions and on many others of the same sort, Locke's positive contributions to the ongoing debates are obvious, as is his engagement with Cartesian ideas. We therefore need to consider carefully what Locke really meant when he said that he was abstaining from physical considerations. He certainly wanted to make it clear that most physical issues cannot be dealt

[29] E II. i. 2.

[30] E I. i. 2. Locke's non-commitment to physics in the *Essay* involves his non-commitment to metaphysics: he does not 'trouble' to elucidate the 'essence' of the mind (ibid.). Interestingly, Locke's agnosticism on this point was the very reason why Condillac, in the Introduction to his *Essay on the Origin of Human Knowledge* (1746), viewed Locke—not Descartes who 'knew neither the origin nor the generation of our ideas'—as the founder of modern *metaphysics*, i.e. the patron of a 'modest' study just of the *operations* of the mind, 'content to stay within the bounds' adjusted to 'the weakness of the human mind'—see the translation and edition of Condillac's *Essay* by Hans Aarsleff, pp. 3–4.

[31] E IV. xxi. 2. A passage in Locke's *Some Thoughts Concerning Education* (1693) reads: 'Natural philosophy being the knowledge of the principles, properties, and operations of *things, as they are in themselves*, I imagine there are two parts of it, one comprehending spirits, with their nature and qualities; and the other bodies' (LW 9, §190, p. 182—our emphasis). Locke points out that natural philosophy requires to be preceded by a study of spirits 'usually referred to metaphysics': 'under what title soever the consideration of spirits comes, I think it ought to go before the study of matter and body, not as a science that can be methodized into a system, and treated of, upon principles of knowledge; but as an enlargement of our minds towards a truer and fuller comprehension of the intellectual world' (§190, p. 183).

with properly until we come to terms with the measure of our own capacity for *knowing*. And as it turns out, most of them cannot be dealt with at all at the level of knowledge, that is, with any hope of achieving certainty about them. Nevertheless, this 'abstentionist' stance on 'physical' issues did not mean that Locke had no opinions, reasoned opinions, on whatever concerns the very being of things. And it did not mean either that Locke's opinions on these matters were not expressed in the *Essay*. Quite the contrary, they are to be found in many places, sometimes in the most explicit way (as in II. viii, where Locke summons up the corpuscular and mechanistic theory of matter to explain the differences between primary and secondary qualities), and sometimes in more cryptic and indirect ways, through suggestions, conjectures, analogies, euphemisms, and so on. In fact, as recent Locke scholarship has shown, and as many contributions here will confirm, the *Essay* as a whole, and Locke's philosophy in general, is offering strong insights into physical and metaphysical issues. And on this ground, the confrontation with Descartes was inevitable. Even though, at the surface of the text, Locke's self-proclaimed agnosticism tended to underplay the importance of these issues, their consideration was still clearly a driving force in the conduct of Locke's arguments.

Summary of the Chapters

J. R. Milton's contribution ('Locke and Descartes: the Initial Exposure, 1660–1670') is a biographical account of Locke's first encounter with Descartes's works. It looks at Locke's manuscript commonplace books with the aim of determining in as much detail as is now possible what books by Descartes Locke read in the period before he started work on the drafts of the *Essay*, what he found of interest in them, and what conclusions might be drawn from this data about his philosophical development. It shows that there is evidence of a considerable and sustained interest in Descartes's mechanical physics but hardly any visible interest in his metaphysics or epistemology—and considers the possible reasons for this state of affairs.

Peter Anstey ('Locke and Cartesian Cosmology') offers more evidence of Locke's interest in Descartes's natural philosophy, and in the Cartesian-inspired scientific literature of the seventeenth century in general. He examines Locke's changing views on the cosmology of Descartes and his followers. In particular, he explains the context in which Locke frames the phrase 'our solar system' and substitutes it for 'our vortex'—a strikingly Cartesian expression which he himself often used, and which was still commonly employed in England at the end of the century.

Chapters 3–5 examine the Cartesian and anti-Cartesian subtext of the Lockean theory of bodies. James Hill's ('The Cartesian Element in Locke's Anti-Cartesian Conception of Body') concentrates on the concept of impenetrability or solidity. It shows that Locke's distinction between hardness and impenetrability parallels the Cartesian one, and it argues that this makes it impossible to ascribe to Locke a strict adherence to the atomistic view, which considers that the indivisibility of the ultimate particles results from their perfect hardness. He makes the case that Locke's agnosticism on the essence of matter is paradoxically derived from the most Cartesian elements in his theory of bodies.

Lisa Downing's 'Are Body and Extension the Same Thing? Locke versus Descartes (versus More)' is also dealing with impenetrability, and shows that Locke's engagement with Descartes goes surprisingly deep on this issue. It illustrates how many of Locke's points on space, extension, and solidity are clarified by seeing them as responding to Descartes's correspondence with More, in which Descartes specifically says that impenetrability results from extension.

Third in this tryptic on the nature of bodies, Martha Bolton's chapter ('Modes and Composite Material Things According to Descartes and Locke') stresses the ontological side of the question. In Descartes's ontology, a created substance, or its principal attribute, unifies the many modes that belong to that substance; by contrast, Locke's ontology includes not only substances and their qualities, but also composite entities which contain substances but are unified by modes. Locke, she argues, seeks to adapt the apparent unity of living things, such as oaks, horses, and human beings, to the (Cartesian) mechanistic doctrine that matter is a substance.

Matthieu Haumesser's chapter ('Virtual Existence of Ideas and Real Existence: Locke's Anti-Cartesian Ontology') considers the concept of 'existence' as it is variously applied in Locke to the objects of sensation (the 'real existence' of things) and to the objects of reflection (the 'fleeting existence' of ideas). It shows that Locke, in order to construct his own ontology and typology of simple ideas and modes, is both using and subverting the Cartesian ontology of substance and modes. Ideas, as 'immediate objects of perception', exist in the mind, but not substantially. This in turn sheds light on the differences between Locke's and Descartes's doctrines of ideas, especially on the question of 'objective reality', which played a strategic part in the *Third Meditation*, as well as in the debate between Arnauld and Malebranche.

Locke's construal of selves, persons, and thinking substances is notoriously difficult and the subject of wide controversy. In Philippe Hamou's chapter ('Locke and Descartes on Selves and Thinking Substances'), it is suggested that we could go some way towards clarifying it by seeing it in the context of Descartes's construal of the same or similar issues. It argues that there are both strong threads of continuity (which may appear even stronger in the light of the recent reappraisal of Descartes's so-called dualism) and a quite obvious (but often neglected) anti-Cartesian strand in Locke's doctrine of the self. The chapter seeks to assess precisely where and why Locke departs from Descartes and shows, contrary to a common but misconceived view of Locke's aim in chapter II. xvii, that it is not so much the Cartesian 'substantiation' of the self that Locke is arguing against, but rather its disembodiment.

Denis Kambouchner ('Locke and Descartes on Free Will') considers interesting parallels in the evolution of Descartes's and Locke's thoughts and formulations with regard to the problem of free will, which, from almost opposite starting points, bring them closer together. The 'family resemblance' between them (also seen in

Malebranche, an important mediator here) is due to the recognition of the irreducibility and complexity of the problem of the determination of the will—a problem that cannot be solved with simplistic formulations such as 'the will is necessitated', or 'the will is absolutely free'. Both Descartes and Locke carefully distinguish between various aspects of the question: whether the will can or cannot be compelled, whether it can resist the attractiveness of certain perceptions, whether the determination of the will obeys rules. When we examine their most carefully considered positions, what appears prima facie as an antinomy between the two doctrines must be significantly nuanced, to the point that the affinities prevail.

Catherine Wilson's chapter ('Essential Religiosity in Descartes and Locke') is a fine example of how broadening our perspective on Locke's relationship to Descartes enables us to better assess the meaning of their epistemological enterprises, and their historical significance. She offers an overview and comparison of Descartes's and Locke's stances toward religious and moral issues (their 'essential religiosity'), such as their views on divine agency in the creation of the world and direction of human affairs; the relevance of divine retribution and reward to morality; their sense of supernatural power and artistry as revealed in things of the world. She also contrasts the different kinds of epistemic and moral humility that these engender in each author.

Laurent Jaffro's chapter ('Locke and Port Royal on Affirmation, Negation, and Other "Postures of the Mind"') claims that in order to properly understand Locke's doctrine of assent, his philosophy of mind needs to be seen in conjunction with his philosophy of language, which in turn gains from being compared with Port-Royal's logic and grammar. He points out two conflicting facts in Locke's account of affirmation and negation in the *Essay*. First, Locke entrusts affirmation and negation with the task of signifying both the assertion by which we manifest our assent to a proposition and the junction or separation of the ideas constituting the proposition. The other fact is that Locke accepts a great variety of ways of considering a proposition. This diversity of 'postures' is poorly expressed by the limited number of syncategorematic terms, 'particles', which he tends to put on an equal footing with the marks of affirmation and negation. The first fact fosters a one-act view of the assent we give to propositions. The second opens the way to a multiple-act view.

Andreas Blank ('Cartesian Logic and Locke's Critique of Maxims') contextualizes Locke's critique of logical and metaphysical maxims within the framework of the Cartesian critique of the topical tradition. It makes clear that Locke, targeting the Scholastic, proof-theoretic conception of maxims, replicates argumentative patterns found in the work of the Cartesian logicians Johannes Clauberg and Antoine Arnauld, who argued against the topical (Ramist) conception of maxims. Locke also inherits certain weaknesses of this Cartesian critique, which, it is argued, does not adequately capture the view of Petrus Ramus and others in the topical tradition that maxims only make explicit the rules that implicitly govern various areas of discourse.

Finally Nicholas Jolley's chapter ('Locke and Malebranche: Intelligibility and Empiricism') addresses the issue of whether Locke's own empiricist theory of ideas offers, as Locke often suggested, a more intelligible way of explaining human understanding than Malebranche's doctrine of Vision in God. Drawing on Locke's statements about the corpuscularian hypothesis, he argues that although the empiricist theory may satisfy some criteria of intelligibility, it is forced to recognize the existence of processes that are 'incomprehensible'; to that extent, Locke's theory of ideas runs parallel with his mature philosophy of matter. The epistemic status of the empiricist theory of ideas is thus more problematic than it is often taken to be.

References

Manuscripts: MS Locke c. 28 fos. 119–20, Oxford, Bodleian Library. The Locke Digital Project <http://www.digitallockeproject.nl>.

Bonno, Gabriel. *Les Relations intellectuelles de Locke avec la France*. Berkeley and Los Angeles: University of California Press, 1955.

Condillac, Étienne Bonnot, abbé de. *Essay on the Origin of Human Knowledge* (1746). Trans. H. Aarsleff. Cambridge: Cambridge University Press, 2001.

De la Motte, Charles. 'La Vie de Coste et anecdotes sur ses ouvrages', in John Locke, *Que la Religion Chrétienne est très-raisonnable*, ed. Hélène Bouchilloux and Maria Cristina Pitassi. Oxford: Voltaire Foundation, 1999: 231–60.

Downing, Lisa. 'Locke and Descartes', *A Companion to Locke*, ed. M. Stuart. Hoboken: John Wiley & Sons, 2015: pp. 100–20.

Gibson, James. *Locke's Theory of Knowledge and its Historical Relations*. Cambridge: Cambridge University Press, 1917, reprinted 2010.

Hatfield, Gary. 'Epistemology and Science in the Image of Modern Philosophy: Rorty on Descartes and Locke', in Juliet Floyd and Sanford Shieh, eds., *Future Pasts: The Analytic Tradition in Twentieth-Century Philosophy*. New York: Oxford University Press, 2001: 393–413.

Jolley, Nicholas. *Locke: His Philosophical Thought*. Oxford: Oxford University Press, 1999.

Jolley, Nicholas. 'Lockean Abstractionism versus Cartesian Nativism', in Paul Hoffman, David Owen, and Gideon Yaffe, eds., *Contemporary Perspectives on Early-Modern Philosophy: Essays in Honor of Vere Chappell*. Peterborough, Ont.: Broadview Press, 2008: 157–71.

Jolley, Nicholas. 'Dull Souls and Beasts: Two Anti-Cartesian Polemics in Locke', in Petr Glombíček and James Hill, eds., *Essays on the Concept of Mind in Early-Modern Philosophy*. Newcastle upon Tyne: Cambridge Scholars, 2010: 97–114.

Lennon, Thomas. *The Battle of the Gods and Giants: The Legacies of Descartes and Gassendi 1655–1715*. Princeton: Princeton University Press, 1993.

Lough, John. *Locke's Travels in France, 1675–1679, as Related in his Journals, Correspondence and Other Papers*. Cambridge: Cambridge University Press, 1953.

Pécharman, Martine. 'Le Problème de la distinction des idées', in P. Hamou and M. de Gaudemar, eds., *Locke et Leibniz: Deux styles de rationalité*. Hildesheim: Georg Olms Verlag, 2011: 13–45.

Rogers, G. A. John. 'Descartes and the Mind of Locke: The Cartesian Impact on Locke's Philosophical Development', in G. A. J. Rogers, *Locke's Enlightenment*. Hildesheim: Georg Olms Verlag, 1998: 23–31.

Rorty, Richard, *Philosophy and the Mirror of Nature*. Princeton: Princeton University Press, 1979.

Schmaltz, Tad M., ed. *Receptions of Descartes: Cartesianism and Anti-cartesianism in Early Modern Europe*. London and New York: Routledge, 2005.

Schouls, Peter A. 'The Cartesian Method of Locke's *Essay concerning Human Understanding*', *Canadian Journal of Philosophy*, 4 (1974/5), 579–601.

Schouls, Peter A. *The Imposition of Method: A Study of Descartes and Locke*. Oxford: Oxford University Press, 1980.

Schuurman, Paul. *Ideas, Mental Faculties, and Method: The Logic of Ideas of Descartes and Locke and its Reception in the Dutch Republic, 1630–1750*. Leiden: Brill, 2004.

Schuurman, Paul. 'Descartes, René (1596–1650)', in S.-J. Savonius-Wroth, Paul Schuurman, Jonathan Walmsley, eds., *The Continuum Companion to Locke*. London: Continuum Press, 2010: 53–7.

Thilly, Frank. 'Locke's Relation to Descartes', *The Philosophical Review*, vol. 9, no. 6 (1900), 597–612.

Yolton, John. *Locke and the Way of Ideas*. New York: Oxford University Press, 1956.

1

Locke and Descartes

The Initial Exposure, 1658–1671

J. R. Milton

Although it is clear from Locke's published writings that he had at least a broad understanding of Descartes's philosophy, he told us nothing specific about how, and when, this knowledge was acquired, or when he first began reading Descartes's works. In 1697 he robustly denied Edward Stillingfleet's claim that he had taken parts of his philosophy from Descartes, but freely admitted that he 'must always acknowledge to that justly admired Gentleman, the great Obligation of my first Deliverance from the unintelligible way of talking of the Philosophy in use in the Schools in his time';[1] though this is regrettably imprecise, it does at least indicate that the encounter was fairly early in Locke's life. One of his closest friends was able to add a little more detail: in January 1705, less than three months after Locke's death, Lady Masham wrote a letter to Jean Le Clerc that included the following much-quoted passage:

> The first Books (as Mr Locke himself has told me) which gave him a relish of Philosophical studys were those of Descartes. He was rejoyced in reading of these because tho' he very often differ'd in Opinion from this writer, he yet found that what he said was very intelligible: from whence he was incourag'd to think That his not haveing understood others, had, possibly, not proceeded altogether from a defect in his Understanding.[2]

No date was given for when this happened, but the language used and the position of these remarks in her account suggest that she put it some years after Locke arrived at Oxford.

Masham's report was incorporated into Le Clerc's own account of Locke's life, from which it passed in truncated form into the short biography appended to editions of his collected works from the fifth edition onwards.[3] It has been quoted by all Locke's main

[1] *A Letter to Edward Lord Bishop of Worcester*, p. 103 (LW 4, p. 48).

[2] Woolhouse, 'Lady Masham's Account of Locke', p. 173; also in Le Clerc, *Epistolario*, 2, p. 500.

[3] 'Eloge de feu Mr Locke', p. 349; *The Works of John Locke, Esq.* (London, 1751), vol. 1, p. v.

biographers. H. R. Fox Bourne did not make a definite estimate of when he supposed that these events happened (though he described them in a chapter covering the years between 1652 and 1660), but in his account of the origins of the *Essay* he stated that Locke first began reading Descartes as an undergraduate, that is, not later than 1656.[4] Maurice Cranston put the encounter much later, about 1666.[5] Roger Woolhouse, relying on an earlier study by the present author, put it in the early 1660s.[6]

Locke's Library

For further information we need to consult Locke's private papers, and the books in his library. Here the obvious place to start looking is the catalogue produced by John Harrison and Peter Laslett. *The Library of John Locke* is an invaluable resource for Locke scholars, but it does have two serious drawbacks. One is quite undisguised: nothing at all is said about *when* any of the volumes listed came into Locke's possession, or when he read any of them.[7] The other is that what Harrison and Laslett produced is neither a simple edition of Locke's own catalogue[8] nor an attempt to assemble a new catalogue *ab initio* from all the evidence available, but rather a somewhat ungainly hybrid of the two. The problems that this way of proceeding can sometimes raise may be seen from the entries on Descartes in the form in which Harrison and Laslett gave them:

CARTES, René des

601^{a}. Opera philosophica . . . 3^{a} ed. . . . 5 pts. 4º, Amstelodami, 1658. $^{8}_{1}$.
Oak Spring.[9]

602. Meditationes de prima philosophia . . . 4º, Amstelodami, 1658.
143[10]. Lo: 4º Am. $\underline{58}$. p. $\overline{226}$ L. *Forms pt. of* 601^{a}.

603. Appendix continens objectiones quintas & septimas in Renati Des-Cartes Meditationes de prima philosophia. 4º, Amstelodami, 1657.
143 Lo: Am: $\underline{57}$. p. $\overline{199}$. $^{8}_{1}$ L.

[4] *The Life of John Locke*, 1, p. 62, 2, p. 89.

[5] *John Locke*, p. 100.

[6] *Locke: A Biography*, p. 35, citing Milton, 'Locke at Oxford'.

[7] The one partial exception is Appendix I (LL, pp. 269–77), which contains the list Locke made in his journal on 14 July 1681 of books in his rooms at Christ Church: Bodleian Library [Bodl.], MS Locke f. 5, pp. 92–103.

[8] Strictly speaking there are three separate catalogues: Bodl., MS Locke f. 16, MS Locke e. 3, and Locke's interleaved copy (LL 1539^{a}) of the *Catalogus Impressorum Librorum Bibliothecae Bodlejanae in Academia Oxoniensi* (Oxford, 1674).

[9] The references to 'Oak Spring' are to the library of Paul Mellon, who was then the owner of these books; they are now in the Bodleian Library. The superscript letter in the number 601^{a} is an indication that this volume (as distinct from its component parts) was not listed separately in Locke's copy of the 1674 Bodleian catalogue (Bodl., Locke 17.16).

[10] This is the page number of the entries on Descartes in the Bodleian catalogue.

Another entry. Appendix continens objectiones 5as & 7as in Renati des Cartes meditationes de prima philosophia 4° Am: $\underline{57}$. p. $\overline{199}$. $\frac{8}{1}$. *Forms pt. of* 601^{a}.

604. Cartes, Ren. Epistolæ pars 1^{a}. 4° Am: [16]$\underline{68}$. p. $\overline{383}\,\frac{9}{102}$ pars 2^{a}. ib. p. $\overline{404}$. Oak Spring.

605. Epist: ad Gis: Voetium. ib. p. $\overline{108}\,\frac{8}{1}$. *Forms pt. of* 601^{a}.

606. Principia philosophiæ. 4° Am. [16]$\underline{56}$. p. $\overline{222}\,\frac{8}{1}$. *Forms pt. of* 601^{a}.

607. [Specimina philosophiæ: seu] Dissertatio de methodo ib. p. $\overline{247}\,\frac{8}{1}$. *Forms pt. of* 601^{a}.

608. Passiones animæ ib. p. $\overline{92}$. $\frac{8}{1}$. *Forms pt. of* 601^{a}.[11]

Another work was not listed by Locke under Descartes's name, though he could hardly have been unaware of its authorship:

2451. A discourse of a Method for the well guiding of Reason & y^{e} discovery of Truth in the sciences 8° Lon [16]$\underline{49}$. $\overline{127}\,\frac{7}{455}$.

This was an English translation of the *Discours de la méthode*; no author was indicated on its title-page, though the translator's preface made it clear who had written it.

It emerges from the entries just described that Locke owned three volumes containing works by Descartes: the translation of the *Discours* just mentioned, the Latin version of the first two parts of Clerselier's edition of Descartes's letters (item 604), and—most importantly—a fat quarto volume containing the Latin versions of Descartes's main philosophical works (item 601^{a}). It is no accident that all these volumes are in either Latin or English. Locke only started seriously to learn French once he had arrived in Montpellier at the start of 1676;[12] there is no sign anywhere in his papers that he bought anything in French before this, and there are no quotations from any French-language works in any of his pre-1675 notebooks.

It is tempting to believe that Locke acquired the English translation of the *Discours* when he was quite young, but the earliest mention of it anywhere among his papers is in the list James Tyrrell made of the books he had been looking after while Locke was in exile in the Netherlands.[13] This indicates that Locke must have acquired his copy by 1683, but beyond that one can only speculate.

Rather more can be said about the 1668 edition of Descartes's letters. An entry in Locke's 1669 memorandum book shows that he bought a copy on 4 August,[14] and if this was the volume listed in his library catalogues—as would seem very likely—it contained both the first and second parts, the only two which had then been published.[15]

[11] LL 101. The list continues with a book (item 609) about Descartes rather than by him: Gabriel Daniel's *Voiage du monde de Descartes* (Paris, 1691).

[12] Journal 9 January 1676, Bodl., MS Locke f. 1, p. 37.

[13] Bodl., MS Locke f. 17, fo. 45^{v}.

[14] British Library [BL], Add. MS 46470, fo. 28^{r}; the price was 12 shillings.

[15] Bodl., Locke 9.102. The third part was published in Amsterdam in 1683; there is no evidence that Locke ever owned a copy.

It was among the books kept in his rooms at Christ Church in 1681, and it too was subsequently looked after by Tyrrell.[16]

By far the most important of these three items is, however, number 601[a], the volume of philosophical writings. What did it actually contain? At first sight the answer seems to be given by the entries that follow it in Harrison and Laslett's catalogue, or—to be more precise—by items 602, 603, and 605 to 608. It might appear from these that the volume contained only the fifth and seventh sets of the *Objections and Replies* (item 603), and not any of the others, but this is not the case. If the volume itself[17] is inspected it immediately becomes apparent that it contains them all: the first four and the sixth sets are in the part that contains the *Meditations* (item 602).

When Locke acquired this volume is uncertain: it was among the books at Christ Church in 1681 and was kept safe for him by Tyrrell while he was in the Netherlands,[18] but there appears to be no record of its purchase anywhere in his papers. However, the fact that the large collection of works on various aspects of post-Cartesian philosophy which Locke brought back from France in 1679 included nothing by Descartes himself suggests that he already possessed a copy of his main works; if he did, it was presumably this one.[19]

Locke's Commonplace Books

A survey of Locke's library reveals which works he had thought it worth buying, but unless the volumes contain marginal annotations—and none of those by Descartes do—they can tell us nothing about what particular topics attracted his attention. For information about this we need to turn to the commonplace books into which he copied extracts from the books he had been reading. It is, of course, extremely unlikely that these record everything—or even nearly everything—that he read, but they do contain quite enough for the direction of his interests to be ascertained, especially for the years he spent in Oxford between 1658 and 1667.[20]

Unfortunately anyone wishing to explore this material faces an immediate problem: the entries in Locke's commonplace books made before 1679—and these are the large majority—are undated. Fortunately, with a certain amount of work it becomes possible to give fairly precise dates to nearly all this material. The reason why this can be done is that from about 1660 onwards Locke organized the entries in his commonplace

[16] Bodl., MSS Locke f. 5, p. 92; f. 17, fo. 28v; b. 2, fos. 121r, 123r.

[17] Bodl., Locke 8.1.

[18] Bodl., MSS Locke f. 5, p. 94; f. 17, fo. 28v; b. 2, fos. 121r, 123r.

[19] On books Locke acquired while in France, see Lough, 'Locke's Reading during his Stay in France'. A nearly complete collection of Descartes's works in French (10 volumes in all) that was listed in Locke's journal on 8 August 1677 (MS Locke f. 2, pp. 226–7) was bought for his pupil Caleb Banks: see Bodl., MS Locke b. 2, fos. 5r, 6r. There are no records of Locke buying anything by Descartes for himself.

[20] Whether Locke actually did more reading during these years or was merely more assiduous in taking notes on what he read is something about which one can only speculate, but there are some reasons to suspect the former.

books according to the method that he was to describe to the world—or at least to readers of the *Bibliothèque universelle*—in 1686. Each entry, whether a quotation from a book or (more rarely) a comment made by Locke himself, was tagged by a title[21] of one or (very occasionally) two words, and then allocated a place in the book determined by the first letter and next vowel of this title, so that 'Motus' would be filed under 'Mo', 'Aqua' and 'Aurum' under 'Au', and so on.[22] Each two-letter class was allocated a pair of facing pages, and when (as often happened) these pages became full, Locke would turn forward until he came to the first pair of entirely blank pages, and continue writing at the top of the left-hand page. These classes are therefore not in any fixed order: to give examples from two notebooks into which he copied material from Descartes, Additional Manuscript 32254 in the British Library begins with the sequence 'Fe', 'Ve', 'Hi', 'Ao', and 'Ba', while MS Locke f. 14 in the Bodleian starts with 'Co', 'Gu', 'Ca', 'Hi', and 'Gi'.

In any notebook all the entries that fall under any class are in strict chronological order, as are the *first* entries on successive left-hand pages, and these two facts make it possible to assemble a consolidated list that gives the approximate order in which the books cited in the notebook were read.[23] Once this has been done, one can then employ a kind of stratigraphical method. Just as archaeologists can often date the strata they unearth by means of coins minted at a known date—or at least during a known interval—so the entries in Locke's commonplace books can be dated by means of the publication dates of the works cited. The way this can be done may be seen in MS Locke f. 14. This is a small notebook that Locke began using in 1659 or thereabouts, and finished using in 1666 or 1667.[24] It is unique among Locke's commonplace books in that it was used for recording comments made about various authors (rather than subjects), with the entries tagged by their names, and one result of this is that it appears to cover the whole range of Locke's reading during that time (though not, of course, mentioning everything that he read). The entries are arranged according to Locke's standard system for organizing a commonplace book, apart from those in the first twenty pages which had been made before he started using this. If one employs the kind of archaeological method just described, it can be seen that the lowest stratum of entries contains nothing published after 1659, but as one moves upwards works from the 1660s begin to appear: first from 1660, then from 1661, and so on. The following

[21] In the English translation (not made by Locke) of the 'Méthode nouvelle de dresser des recueuils' in the 1706 *Posthumous Works* these are referred to as 'heads', but this was not Locke's own term: in the 'Méthode nouvelle' they are 'titres', and in Locke's autograph of an earlier English version he called them 'titles': BL, Add. MS 28728, fo. 58^r.

[22] The titles were always in Latin, whatever the language of the entry that followed.

[23] Though when only a very few quotations were taken from any book, a rough dating is often all that is possible; entries made at or near the end of any class are also often hard to date precisely.

[24] A mistaken date of 1667 for all the entries in this notebook is one of the main weaknesses of what is otherwise still the most thorough study of Locke's familiarity with Cartesianism: Charlotte S. Ware's 'The Influence of Descartes on John Locke'. On its organization and date, see Milton, 'The Date and Significance of Two of Locke's Early Manuscripts'.

(slightly simplified[25]) list of the books cited in this manuscript, in the order in which they were read, should make clear:

C. V. Schneider, *Liber de Osse Cribriformi* (Wittenberg, 1655), LL 2580.
Sir George Ent, *Apologia pro Circulatione Sanguinis* (London, 1641), LL 1054.
Nathaniel Highmore, *Corporis Humani Disquisitio Anatomica* (The Hague, 1651), LL 1451[a].
E. J. Brochmand, *Ethices Historiae Specimen* (Leiden, 1653).

Robert Boyle, *New Experiments Physico-Mechanicall, Touching the Spring of the Air* (Oxford, 1660).
Jean-Louis Guez de Balzac, *Aristippus. Or, Mon[sr] de Balsac's Masterpiece* (London, 1659), LL 185.
Pierre Gassendi, *Opera Omnia In Sex Tomos Divisa* (Lyon, 1658).
René Descartes, *Principia Philosophiae* (Amsterdam, 1656).
Jean-Louis Guez de Balzac, *The Prince* (London, 1648), LL 182.

Joseph Glanvill, *The Vanity of Dogmatizing* (London, 1661).
Robert Boyle, *Certain Physiological Essays* (London, 1661), LL 439.
G. J. Vossius, *Chronologiae Sacrae Isagoge* (The Hague, 1659).
Isaac Vossius, *De Septuaginta Interpretibus, eorumque Tralatione & Chronologia Dissertationes* (The Hague, 1661).
Robert Boyle, *Some Considerations touching the Style of the Holy Scriptures* (London, 1661).
Robert Sharrock, *The History of the Propagation & Improvement of Vegetables* (Oxford, 1660).
Lucius Cary, Viscount Falkland, *His Discourse of Infallibility* (London, 1651).
Isaac Vossius, *Dissertatio de Vera Aetate Mundi* (The Hague, 1659).
Hugo Grotius, *De Veritate Religionis Christianae* (Oxford, 1660).
Gabriel Naudé, *Instructions concerning Erecting a Library* (London, 1661).
Athanasius Kircher, *Itinerarium Exstaticum* (Rome, 1656).
Hermann Conring, *De Calido Innato Sive Igne Animali* (Helmstadt, 1647).

J. V. C. [John Vincent Canes], *Fiat Lux* (n.p., 1662).
Girolamo Cardano, *Arcana Politica Sive de Prudentia civili* (Leiden, 1635), LL 590[a].
Jean-Louis Guez de Balzac, *The Choyce Letters of Monsieur De Balzac* (London, 1658).
Adam Olearius, *The Voyages & Travels of the Ambassadors* (London, 1662), LL 2128.
Marten Schoock, *Liber de Cervisia* (Groningen, 1661), LL 2581.
Gaspar Schott, *Magia Universalis Naturae et Artis* (Würzburg, 1657).
Gaspar Schott, *Mechanica Hydraulico-Pneumatica* (Würzburg, 1657).
Anton Deusing, *Considerationes circa Experimenta Physico-Mechanica* (Groningen, 1662).

[25] A small number of infrequently quoted books whose position in the sequence cannot be reliably determined have been omitted.

Antonio Neri, *The Art of Glass* (London, 1662).
John Smith, *Select Discourses* (London, 1660).
Isaac Vossius, *De Lucis Natura et Proprietate* (Amsterdam, 1662).

[Thomas White], *Sciri, Sive Sceptices & Scepticorum à Jure Disputationis Exclusio* (London, 1663).
J. C. Magnen, *Exercitationes de Tabaco* (n.p., 1658), LL 1869.
Robert Boyle, *Some Considerations touching the Usefulnesse of Experimental Naturall Philosophy* (Oxford, 1663), LL 465.

J. E. [John Evelyn], *Sylva, Or A Discourse of Forest-Trees, and the Propagation of Timber* (London, 1664).
Thomas Willis, *Cerebri Anatome* (London, 1664), LL 3165[a].
Benedictus de Spinoza, *Renati Descartes Principiorum Philosophiae Pars I, & II, More Geometrico demonstratae* (Amsterdam, 1663), LL 2742.
Robert Boyle, *The Sceptical Chymist* (London, 1661), LL 444.

J. J. Becher, *Institutiones chimicae prodromae, id est... Oedipus Chimicus* (Amsterdam, 1664), LL 248.
Henry Power, *Experimental Philosophy* (London, 1664).
Pierre Gassendi, *The Mirrour of true Nobility & Gentility being the Life of the Renowned Nicolaus Claudius Fabricius, Lord of Peiresk* (London, 1657).
Thomas Bartholinus, *De Pulmonum Substantia et Motu Diatribe* (Copenhagen, 1663), LL 215.

This list reveals that—as one would expect—many of the books that Locke was reading at any given time had been published earlier, often much earlier; fortunately he seems generally to have been reading enough newly published work for the method described here to work gratifyingly well.

It emerges from these investigations that the pre-1675 quotations from Descartes's writings in Locke's papers were nearly all made in the early 1660s, the only exceptions being those taken from Descartes's letters. Four of these came from the 1668 edition already mentioned—two in the commonplace book known as 'Adversaria 1661', now in private hands,[26] and two in MS Locke d. 11 in the Bodleian Library.

Adversaria 1661 is a large commonplace book that (despite its name) was not used until 1670 or thereabouts;[27] in one of the catalogues of Locke's library it was listed as 'Adversaria Ethica',[28] and this would be a much less misleading title for it than the one by which it is now usually known. It was used not only for records of his reading but also (rather unusually) for copies of several of his works, notably Draft A of the *Essay*. The entries in it are organized according to Locke's standard method.

[26] There is a microfilm in the Bodleian Library, MS Film 77.
[27] For an explanation of this, see Milton, 'The Dating of "Adversaria 1661"'.
[28] LL 23[a]. It was described by Harrison and Laslett as 'Not located', but there can be no doubt whatever about its identity.

MS Locke d. 11 also had a name, 'Lemmata Physica',[29] and like its companion volume 'Lemmata Ethica' (MS Locke d. 10) was organized according to a rather cumbersome system that Locke never used again. Each verso page[30] was dedicated to a two-letter class, but in these two volumes these classes are arranged in alphabetical order, so that MS Locke d. 11 starts with 'Aa' on folio 1^{v}, then 'Ae' on folio 2^{v}, and so on. Each page was then divided into five vertical columns, which were allocated in turn to the next subsequent vowels, so that an entry tagged by 'Fluiditas' was put into the third column of page 'Fu' (Fui), and one on 'Ventus' into the fifth column of 'Ve' (Veu). This way of working has the disadvantage for us—not for Locke—that individual entries are generally difficult to date with any precision,[31] unless the works quoted happen also to have been cited in other manuscripts, as fortunately they often were.

Descartes's Correspondence

We know from the entry in Locke's 1669 memorandum book that he bought his copy of the 1668 edition of Descartes's letters in August of that year, but the quotations from it in his commonplace books seem to have been copied rather later, probably in 1671.[32] These entries illustrate the difficulties that can arise if one tries using Locke's commonplace books for insights into what he was thinking. The two in Adversaria 1661 both come from an adulatory letter that Descartes wrote to Queen Christina in November 1647,[33] and are on ethical topics: Goodness and Happiness. The two in MS Locke d. 11 are from a rather earlier letter to Princess Elisabeth, and describe Descartes's low opinion of his troublesome disciple Hendrik de Roy (Regius) and his more favourable one of Cornelius van Hogeland.[34] It is clear that Locke had sufficient interest in Descartes to obtain and start reading a volume of his letters, but if one tries going beyond this rather obvious inference, uncertainties close in. Did he read more than the two letters he quoted? Presumably he did, but if so, which ones, and what did he find? We do not know, and unless more evidence shows up, there is not much more that can be said.

[29] Locke started using this volume in 1659, and took it with him when he moved to London in 1667. It remained in fairly frequent use until 1675, but after that was not used again until 1693, when he had settled at Oates.

[30] The recto pages opposite were initially blank (as most of them still are), and were used on those fairly infrequent occasions when a column begun on the facing page had become full.

[31] This is partly because the classes are much more numerous, and the number of entries in each correspondingly shorter, but mainly because the order of the classes is fixed.

[32] The entry on Regius in MS Locke d. 11 (fos. 73^{v}–74^{r}) cannot be earlier than 1671, since it occurs directly below one giving the title of Francesco Redi's *Experimenta circa generationem Insectorum* (Amsterdam, 1671), LL 2453. Those in Adversaria 1661 are definitely earlier than the copy of Draft A of the *Essay* in that manuscript, but probably not very much earlier.

[33] Descartes to Christina, 20 November 1647, AT 5, pp. 81–6.

[34] Descartes to Elisabeth, March 1647, AT 4, pp. 624–8.

There is one other quotation from Descartes's correspondence that seems to be the earliest mention of his name anywhere among Locke's papers. Unlike those just described it did not come from Clerselier's edition: Locke found it in Pierre Borel's short life of Descartes, the *Vitae Renati Cartesii, Summi Philosophi Compendium* published in 1656. The quotation was copied into a very early medical notebook, MS Locke e. 4, and though the entries in this are not at all easy to date, it is likely that this one was made in 1658 or 1659.[35] It is a short note on a letter that Descartes had written to Mersenne in October 1642 on the subject of smoking chimneys and how to deal with them.[36] It is difficult to say what conclusions might be drawn from this, but it hardly provides evidence of any interest in Cartesian metaphysics.

The *Opera Philosophica*

The results of this survey of the citations taken from Descartes's correspondence are dispiritingly meagre. Fortunately the second group of quotations is much more extensive and correspondingly more informative. These are to be found in three main places. More than half (38) are in the commonplace book Additional MS 32554 in the British Library, and almost all the remainder are in two commonplace books in the Bodleian, MS Locke d. 11 (13) and MS Locke f. 14 (7). There is also a solitary one copied into one of Locke's interleaved bibles, also in the Bodleian Library.[37]

MS Locke d. 11 and MS Locke f. 14 have been described already. Additional MS 32554 is another small notebook, though slightly larger than MS Locke f. 14. It is one of the many commonplace books Locke used mainly for entries on medicine, but there are also a considerable number on various aspects of natural philosophy. All the entries are arranged according to Locke's standard system, and while most of them were made in 1660 or 1661, there are quite a few from the years that followed. In both this and MS Locke f. 14 the latest books cited were published in 1666, and it would seem likely that few if any of the entries in either notebook were made after Locke moved to Exeter House in May 1667.

A complete list of the entries citing Descartes's works in these manuscripts is as follows:[38]

Principia Philosophiae [read c.1660]

ii. §13, p. 32	Locus	Add. MS 32554, p. 182
ii. §24, p. 36	Motus	Add. MS 32554, p. 82
ii. §54, p. 50	Dura et fluida	Add. MS 32554, p. 128
ii. §§56, 57[39]	Fluiditas	MS Locke d. 11, fo. 30^{v}

[35] Bodl., MS Locke e. 4, p. 109, quoting *Vitae Renati Cartesii*, 38–9. On the problems involved in dating entries in this notebook, see Milton, 'John Locke's Medical Notebooks', pp. 148–50.

[36] Descartes to Mersenne, 20 October 1642, AT 3, pp. 587–90.

[37] Bodl., Locke 16.25, LL 309.

[38] The Latin titles (except those placed in square brackets) are as Locke gave them.

[39] No page numbers were given.

iii. §35, p. 67	Scheinerus	MS Locke f. 14, p. 86
iii. §35, p. 67	Solis maculae	MS Locke d. 11, fo. 80v
iii. §63, p. 83	Lux	Add. MS 32554, p. 77
iii. §122, p. 126	Soliditas	Add. MS 32554, pp. 162–3
iii. §132, p. 135	Nigredo	Add. MS 32554, p. 186
iv. §16, p. 153	Pelluciditas	Add. MS 32554, pp. 150–1
iv. §18, p. 154	Liquor	Add. MS 32554, p. 17
iv. §20, p. 155	Gravitas	Add. MS 32554, p. 22
iv. §31, p. 160	Rarefactio	Add. MS 32554, p. 184
iv. §45, p. 169	Aer	Add. MS 32554, p. 13
iv. §48, p. 170	Aqua	Add. MS 32554, p. 35
iv. §49, p. 170	Mare	Add. MS 32554, p. 36
iv. §58, p. 176	Sal	MS Locke d. 11, fo. 77v
iv. §64, p. 178	Fons	MS Locke d. 11, fo. 29v
iv. §70, p. 180	Spiritus	Add. MS 32554, p. 86
iv. §78, p. 182	Terrae motus	Add. MS 32554, p. 104
iv. §80, p. 182	Ignis	Add. MS 32554, p. 134
iv. §110, p. 193	Nitrum	Add. MS 32554, p. 186
iv. §118, p. 196	Liquiditas	Add. MS 32554, p. 17
iv. §119, p. 196	Siccitas	Add. MS 32554, p. 86
iv. §124, p. 198	Vitrum	MS Locke d. 11, fo. 89v
iv. §124, p. 198	Vitrum	Add. MS 32554, p. 33
iv. §127, p. 199	Mollities	Add. MS 32554, p. 82
iv. §130, p. 200	Pelluciditas	Add. MS 32554, pp. 150–1
iv. §132, p. 201	Elasticus motus	Add. MS 32554, p. 136
iv. §133, p. 202	Magnes	MS Locke d. 11, fo. 51v
iv. §166, p. 220	Magnes	MS Locke d. 11, fo. 51v
iv. §166, p. 220	Gilbertus	MS Locke f. 14, p. 25
iv. §187, p. 230	Qualitates occultae	Add. MS 32554, p. 68
iv. §191, p. 233	Sensus	MS Locke d. 11, fo. 78v
Dioptrice [read *c.*1660]		
c. 1 n. 3, p. 63	Lumen	Add. MS 32554, p. 76
[c. 1] n. 8, p. 66	Lux	Add. MS 32554, p. 76
c. 1 n. 9, p. 68	Color	Add. MS 32554, pp. 176–7
c. 2	Refractio	Add. MS 32554, p. 90
c. 4 n. 2, p. 82	Nervus	Add. MS 32554, p. 49
Cartes: Op: Diop.	Visus	MS Locke d. 11, fo. 89v
Meteora [read *c.*1660]		
c. 1, p. 184	Calor	Add. MS 32554, p. 188
c. 1 n. 8, p. 186	Siccitas	Add. MS 32554, p. 86
c. 1 n. 8	Aqua	Add. MS 32554, p. 35
c. 1 n. 9	Aqua	Add. MS 32554, p. 35
c. 2 n. 4, p. 190	Calor	Add. MS 32554, p. 188
c. 2 n. 5	Calor	Add. MS 32554, p. 188
c. 3	Mare	MS Locke d. 11, fo. 51v
c. 3	Sal	MS Locke d. 11, fo. 77v
c. 4, p. 206	Ventus	MS Locke d. 11, fo. 88v

c. 4 n. 1, p. 206	Ventus	Add. MS 32554, p. 173
c. 4 n. 3, p. 208	Rarefactio	Add. MS 32554, p. 184
Dissertatio de Methodo [probably read *c.*1660]		
p. 43[40]	Brutum	MS Locke d. 11, fo. 10^{v}
Passiones Animae [probably read *c.*1661]		
Epist. ad Cartesium	Gilbertus	MS Locke f. 14, p. 25
Epist. ad Cartesium	Baco	MS Locke f. 14, p. 92
Meditationes de Prima Philosophia [probably read *c.*1660]		
Meditation 2, p. 11	Corpus	Add. MS 32554, p. 177
Objectiones et Responsiones [read *c.*1662]		
Obj. 5^{ae} p. 126[41]	Augustinus	MS Locke f. 14, p. 79
[Obj. 5^{ae}] p. 136	[Augustinus]	MS Locke f. 14, p. 79
Resp. 5^{ae} p. 66	Mersennus	MS Locke f. 14, p. 64
Resp. 6^{ae} p. 187	[Knowledge of God]	Bible, LL 309, at 1 Cor. viii. 2
Epistolae, Pars Prima (Amsterdam, 1668) [read *c.*1671]		
Epist. 1	Felicitas	Adversaria 1661, p. 1
Epist. 1	Bonum	Adversaria 1661, p. 2
Epist. 22	Hogeland	MS Locke d. 11, fo. 39^{v}
Epist. 22	Regius	MS Locke d. 11, fos 73^{v}–74^{r}

One thing that is immediately apparent from this list is that the overwhelming majority of the entries are on topics in natural philosophy. The most frequently cited work is the *Principles of Philosophy*, and here the distribution of the quotations is far from uniform: there are twenty-five citations of passages in Part 4, on terrestrial physics, in comparison with only five on the celestial physics in Part 3, four on the theoretical physics in Part 2, and none at all on the metaphysics in Part 1. The next most frequently cited work is the *Meteors*, with eleven citations, followed by the *Dioptrics* with six. There is only one citation from the *Meditations* and one from the *Discourse on Method*, concerned respectively with the essence of body and animal automatism.

The entries taken from the *Principles*, the *Meteors*, and the *Dioptrics* seem to have been made within a fairly short time of one another—probably not more than two or three months, and perhaps less. These entries cannot be earlier than the late summer of 1660, since they were made after Locke had read Boyle's *New Experiments Physico-Mechanicall touching the Spring of the Air*, which went on sale in August, but they are unlikely to be very much later than this since they were made before he read Joseph

[40] There is nothing on this on p. 43 of either of the editions published in 1656. The passage apparently being referred to (it is not quoted exactly) is on p. 45 of the Janszoon edition and p. 36 of the Elzevir edition, and it would therefore seem more likely that the former was used.

[41] This is a mistake on Locke's part: the passage quoted actually comes from Arnauld's Fourth set of Objections, where it is on p. 126 of the 1658 Janszoon edition (and p. 108 of the 1654 Elzevir edition).

Glanvill's *Vanity of Dogmatizing* and Boyle's *Certain Physiological Essays*, both of which were published in March 1661 and appear to have been read soon afterwards.[42]

One work that seems to have been read a little later—probably in 1661—is the *Passiones Animae*, the Latin translation of the *Passions de l'âme*. It might easily be supposed that this shows that Locke had become interested in the analysis of the passions, but there are reasons for being cautious: the only two passages quoted come not from the work itself, but from the first of the two anonymous letters to Descartes that were placed at the start of the book, and these were concerned with general issues raised by Descartes's method in natural philosophy. How much of the main part of the book Locke actually read can therefore only be a matter of conjecture, but both the quotations copied into MS Locke f. 14 were comments on writers who had contributed to natural philosophy—Francis Bacon and William Gilbert—rather than to the theory of the passions.

Another work that was read around 1662 was the *Objections and Replies* to the *Meditations*. Apart from one long quotation copied into Locke's large interleaved Bible,[43] all the citations from this are in MS Locke f. 14, the authors chosen for comment being Mersenne and St Augustine.

Locke's Copy of the *Opera Philosophica*

None of the citations in any of these three notebooks gives any direct indication of which edition—or editions—of Descartes's works Locke was using, but at the back of Additional MS 32554 there is a long list of the works that he took excerpts from when he first started using the notebook,[44] the last one being 'Cartesius Opera 4° Amst[erdam] 1656'. This sounds like the volume listed as item 601[a] by Harrison and Laslett, but closer enquiry shows that it cannot have been.

There were in fact two quite distinct Latin editions of Descartes's philosophical writings produced in the mid-1650s, one (with dates on their title-pages between 1654 and 1656) by the Elzevirs—Descartes's original publisher for these works—and the other (with dates from 1656 to 1658) by another Amsterdam bookseller, Jan Janszoon.[45] The titles of the volumes they produced are as follows:

[42] On the works by Boyle, see *The Works of Robert Boyle*, 1, p. cxxxi; II, p. xiv. The dedicatory epistle in the *Vanity of Dogmatizing* was dated 1 March 1660 (i.e. 1661).

[43] This cannot be dated with any precision, but it would seem likely to have been made around the same time as the entries derived from the *Objections and Replies* in MS Locke f. 14.

[44] BL, Add. MS 32554, pp. 244–5. There are thirty-eight works in this list, of which three were published in 1660 (and none later): Thomas Willis, *Diatribae Duae Medico-Philosophicae* (London, 1660), Pietro Maria Canepari, *De Atramentis* (London, 1660), and Nathaniel Highmore, *Exercitationes Duae* (Oxford, 1660). According to notes made by the London book-collector George Thomason on the title-pages of the copies he bought, he acquired both *De Atramentis* (BL, E.1042[1]) and *Exercitationes Duae* (BL, E.1968[1]) in September.

[45] Guibert, *Bibliographie des oeuvres de René Descartes publiées au XVIIe siècle*, 50–5, 88–90, 105–7, 124–6, 160–2.

1E *Renati Des Cartes Meditationes De Prima Philosophia, In quibus Dei existentia, & animæ humanæ à corpore distinctio, demonstrantur. His adjunctæ sunt variae objectiones doctorum virorum in istas de Deo & anima demonstrationes; Cvm Responsionibvs Avthoris. Editio ultima prioribus auctior & emendatior.*	**1J** *Renati Des Cartes Meditationes De Prima Philosophia, In quibus Dei existentia, & animæ humanæ à corpore distinctio, demonstrantur. His adjunctæ sunt variae objectiones doctorum virorum in istas de Deo & anima demonstrationes; Cum Responsionibus Authoris, Editio ultima prioribus auctior & emendatior.*
Amstelodami, Apud Ludovicum Elzevirium. M DC LIV.	Amstelodami, Apud Johannem Janssonium Juniorem, Anno M. DC. LVIII.
2E *Appendix, Continens Obiectiones Quintas & Septimas In Renati Des-Cartes Meditationes De Primâ Philosophiâ, Cum ejusdem ad illas Responsionibus & duabus Epistolis, Vna ad Patrem Dinet Societatis Iesu Præpositum Provincialem per Franciam, Altera ad celeberrimum Virum D. Gisbertvm Voetium.*	**2J** *Appendix, Continens Obiectiones Quintas & Septimas In Renati Des-Cartes Meditationes De Primâ Philosophiâ, Cum ejusdem ad illas Responsionibus & duabus Epistolis, Vna ad Patrem Dinet Societatis Iesu Præpositum Provincialem per Franciam, Altera ad celeberrimum Virum D. Gisbertum Voetium.*
Amstelodami, Apud Ludovicum Elzevirium, M DC LIV.	Amstelodami, Apud Johannem Jansonium Juniorem, M DC LVII.
3E *Renati Des-Cartes Principia Philosophiæ. Nunc demum hac Editione diligenter recognita, & medis expurgata.*	**3J** *Renati Des-Cartes Principia Philosophiæ.*
Amstelodami, Apud Ludovicum & Danielem Elzevirios, Anno M DC LVI.	Amstelodami, Apud Johannem Jansonium Juniorem, Anno M DC LVI.
4E *Renati Des Cartes Specimina Philosophiæ: sev Dissertatio de Methodo Rectè regendæ rationis, & veritatis in scientiis investigandæ: Dioptrice, et Meteora. Ex Gallico translata, & ab Auctore perlecta, variisque in locis emendata. Nunc denuo hac Editione diligenter recognita, & medis expurgata.*	**4J** *Renati Des Cartes Specimina Philosophiæ: seu Dissertatio de Methodo Rectè regendæ rationis, & veritatis in scientiis investigandæ: Dioptrice, et Meteora. Ex Gallico translata, & ab Auctore perlecta, variisque in locis emendata.*
Amstelodami, Apud Ludovicum & Danielem Elzevirios, Anno M DC LVI.	Amstelodami, Apud Johannem Janssonium Juniorem, Anno M DC LVI.
5E *Passiones Animæ, per Renatvm Des Cartes: Gallicè ab ipso conscriptæ, nunc autem in exterorum gratiam Latina civitate donatæ. Ab H. D. M.* I. V. L.	**5J** *Passiones Animæ, per Renatum Des-Cartes: Gallicè ab ipso conscriptæ, nunc autem in exterorum gratiam Latina civitate donatæ Ab H. D. M.* I. V. L.
Amstelodami, Apud Ludovicum & Danielem Elzevirios, Anno M DC LVI.	Amstelodami, Apud Johannem Janssonium Juniorem. Anno M DC LVI.

It should be emphasized that what we have here is not the (quite common) situation of sheets printed for an older edition being reissued by another publisher with a new title-page. All these are entirely distinct editions with different numbers of pages in every case, and it is these differences of pagination that make it possible to determine—at least in most cases—which edition Locke was using.

An examination of the volume that Locke owned shows that it is a composite made up of the Janszoon editions of the *Meditations* and the *Objections and Replies* (**1J** and **2J**) together with the Elzevir editions of the other pieces (**3E**, **4E**, and **5E**). The quotations from Descartes's writings in Locke's commonplace books were, however, all taken from the Janszoon editions, with the possible (though doubtful) exception of those from the *Passions of the Soul*, which could have come from either edition.[46]

It is therefore clear that when Locke was reading the *Principles* and the other works on natural philosophy in the early 1660s, he was *not* using the volume of Descartes's philosophical writings that he subsequently owned.[47] There is nothing hugely surprising about this: only a minority of the works that Locke made notes on while he was living at Christ Church in the 1660s were subsequently in his library, and some of these may have been later purchases. In his later years Locke had enough money to buy virtually anything that he wanted to read, but when he was young funds were much tighter, and his book purchases seem to have been correspondingly fewer.

Mentions of Descartes by Other Authors

Locke's commonplace books contain not only extracts from Descartes's works, but also remarks made about him by other writers. These came from a wide range of works: Robert Boyle's *New Experiments Physico-Mechanicall, Touching the Spring of the Air* and *Certain Physiological Essays*, Thomas Willis's *Cerebri Anatome*, an English translation of Balzac's letters,[48] and two works by English Catholic authors: John Vincent Canes's *Fiat Lux* and Thomas White's *Sciri, Sive Sceptices & Scepticorum à Jure Disputationis Exclusio*.[49] Perhaps the most interesting of these entries was taken from Lodewijk Meyer's preface to Spinoza's reworking of Descartes's *Principles*, published in 1663.[50] It is, unfortunately, the only quotation from this book anywhere among Locke's papers, but he seems to have been sufficiently interested in what he

[46] The pages of the *Passiones Animae* containing the prefatory letters were not numbered in either edition, which explains the absence of any page numbers in Locke's citations.

[47] Since the quotations from the *Objections and Replies* in both MS Locke f. 14 and the interleaved Bible came from the Janszoon edition, Locke could perhaps have taken them from the copy that he subsequently owned, if he had bought it by 1662; there is, however, no reason to suppose that he did so, or indeed that he had acquired the volume as early as this.

[48] Jean-Louis Guez, Sieur de Balzac, *The Choyce Letters of Monsieur De Balzac* (London, 1658), LL 184, letter to Valentin Conrart, 24 January 1640.

[49] Bodl., MS Locke f. 14, pp. 22 (Boyle), 138 (Canes), 139 (Balzac, White, and Willis).

[50] *Renati Descartes Principiorum Philosophiae Pars I, & II, More Geometrico demonstratae* (Amsterdam, 1663), LL 2742. The position of the entry in MS Locke f. 14, beginning on p. 139 and concluding on p. 190, shows that it was made in 1664. An earlier entry on p. 139 was taken from Willis's *Cerebri Anatome*, licensed on 20 January 1663[/4], and the first entry on p. 188 was taken from Evelyn's *Sylva*, published in mid-February: *The Diary of John Evelyn*, iii. 368. The entry on p. 190 immediately following the one from Spinoza was taken from the English translation of Gassendi's life of Peiresc, *The Mirrour of true Nobility & Gentility* (London, 1657), which Locke read in the autumn of 1664 (there are numerous quotations in Bodl., MS Locke f. 27, pp. 2–7).

found to make a note in his pocket memorandum book asking what else Spinoza had written.[51]

Conclusion

The picture that emerges from the evidence described here is easy enough to summarize. At some time in the early 1660s—probably in the second half of 1660 or the first half of 1661—Locke made detailed notes on the parts of the *Principles of Philosophy* that dealt with natural philosophy, and on the *Dioptrics* and the *Meteors*; at around this time or soon afterwards he also made solitary notes on the *Discourse on Method* and on the *Meditations*. A year or two later he made a small number of notes on one of the letters prefaced to the *Passions of the Soul*, and on the *Objections and Replies* to the *Meditations*. None of these notes was directly concerned with metaphysical or epistemological issues.

How should this evidence be interpreted? In broad terms there seem to be three possible explanations:

(1) In the early 1660s Locke became deeply interested in the kind of mechanistic physics expounded by Descartes and Boyle, and his note-taking reflected this. He had at this stage no comparable interest in metaphysics, and his note-taking reflected this too.

(2) Locke did not make any notes on metaphysics because he used his commonplace books primarily for records of factual data, and this is not what metaphysics is about.

(3) Locke did take at least some notes on metaphysical matters, but unfortunately these have not been preserved.

It has to be acknowledged that choosing between these three explanations is far from straightforward: all of them 'save the phenomena', in the sense of being entirely compatible with the evidence presented here, and any preference for one of them has to be made on other grounds.

Option (3) opens up a large and difficult problem: how many of Locke's manuscripts have been lost? Sometimes we can give firm answers: we undoubtedly have a complete set of the journals, begun when Locke went to France in November 1675, but in the case of the pocket memorandum books that preceded them, it is clear that we possess some but not others: those for 1668, 1670, 1671, 1673, and 1675 are missing.[52] In the

[51] In a list headed 'Q 64' Locke noted: 'Spinoza | Quid ab eo scriptum praeter partem 1 & 2 principiorum Cartesii 4° 63', Bodl., MS Locke f. 27, p. 166.

[52] The earliest of these memorandum books known to exist (and the earliest to survive) is Bodl., MS Locke f. 27, begun in 1664 and used until early 1666. Another memorandum book was used in late 1666 and 1667; this is now in the National Library of Israel (MS Var. 294), and there is a microfilm in the Bodleian Library, MS Film 79. The 1669 book is in the British Library (Add. MS 46470), and the 1672 and 1674 books in the Bodleian (MSS Locke f. 48 and f. 13). After Locke started keeping his journal in

area most relevant to the present study, namely the commonplace books, we have all the large ones listed in Locke's library catalogues, with the exception of a volume referred to as 'Adversaria 62', which was not mentioned in any of the post-1690 catalogues but which had been listed among his books at Christ Church in 1681.[53] To judge from a handful of cross-references to it found elsewhere in Locke's notebooks, its contents were medical.[54]

Another way of estimating the number of lost commonplace books is by looking at cross-references in Locke's other notebooks, and seeing how many of these point to a volume that still exists, and how many do not. An investigation of this seems to indicate that most—indeed the large majority—do. There is no doubt that some commonplace books have been lost, but as far as can be determined these are only a fairly small fraction of those that survive. (The one exception concerns Locke's very early notebooks—those he used at school and during his first years at Oxford—where the attrition rate seems to have been much higher.[55]) In this connection it is worth mentioning that although Lord King's biography of Locke (first published in 1829 and expanded in 1830) gave extracts from a very wide selection of the papers that were then in his possession, he quoted nothing from any manuscript that has since been lost, apart from a short paper of very uncertain authorship apparently designed for a visitor to England, and one paragraph taken from a draft of a letter.[56]

Option (2) has some general points in its favour.[57] It is striking how little philosophy (in the narrow modern sense) there is in *any* of Locke's commonplace books, even from periods when we know from other evidence that he was hard at work on the *Essay*. Sometimes, as when he was in France or in Holland, this can be explained by the fact that he did not have any of his commonplace books with him (they had all been left behind in England), but even when he was at home in London or Oxford the general picture is the same. Locke must have spent a significant part of 1671 and 1672 writing Drafts A and B of the *Essay*, but the only material connected even indirectly with these labours that surfaces anywhere in his commonplace books—apart from the copy of Draft A itself in Adversaria 1661—are two mentions in that notebook of the 1656 edition of Herbert of Cherbury's *De Veritate*, cited in Draft B.[58]

November 1675 the series ceases, though he did continue to use three small unbound pocket-books for similar purposes: Bodl., MSS Locke f. 15, f. 28, and f. 29.

[53] LL 270; it was described there as a quarto volume, 10 inches tall. The volume 'Adversaria 59' mentioned on the same page is probably MS Locke d. 10. Two other volumes in the same list were entitled 'Institutio' and 'Praxis'; Harrison and Laslett were not able to identify either (LL 214, 269), but in fact they are two large medical notebooks now in the Biblioteca Marciana in Venice, where they have been bound together as MS Lat. VII, p. 22.

[54] Bodl., MS Locke f. 23, pp. 14, 18, 20, 22–4, 26; MS Locke f. 28, p. 127; MS Locke f. 29, p. 28.

[55] The undated (but clearly very early) draft of a letter to G.O.W. (LC 1, pp. 13–14) is on a leaf torn out of a notebook on Greek grammar, Bodl., MS Locke c. 25, fos. 43–4 appears to be an index from another lost notebook, and Locke's copy of Joachim Jungius' *Doxoscopiae Physicae Minores* (Hamburg, 1662), LL 1593, was bound using some early geographical notes as endpapers.

[56] *The Life of John Locke* (London, 1829), 4–5, 133–6; the letter is in LC 4, pp. 15–16.

[57] This is the explanation adopted by G. A. J. Rogers, *Locke's Enlightenment*, 25–6.

[58] Adversaria 1661, pp. 144, 146; Draft B, §6, pp. 111–12.

Despite the attractions of options (2) and (3), my own inclinations are firmly towards option (1). If we attempt to draw a picture of Locke's intellectual agenda in the 1660s, what emerges is someone with diverse but essentially uncoordinated interests: academic medicine, the role of the magistrate in ordering religious worship, corpuscularian natural philosophy, the law of nature, practical chemistry, religious toleration, rates of interest, and so on. There is no sign that he saw these various investigations as forming parts of a larger and more systematic project, in the way that Hobbes so manifestly did with his own enquiries. In the Epistle to the Reader of the *Essay* Locke described how after the initial meeting in his chamber he went away and put down on paper 'Some hasty and undigested Thoughts, on a Subject I had never before considered' (E p. 7). There is no reason to suppose that he was being disingenuous here. If he had said this about natural philosophy he would have been not merely disingenuous but downright dishonest, but (of course) he made no such claim. In the case of an enquiry into the powers and limits of the human understanding, he could quite truthfully have said this, and he did.

References

Borel, Pierre. *Vitae Renati Cartesii, Summi Philosophi Compendium*. Paris: Jean Billaine and Matthieu Dupuis, 1656.

Bourne, H. R. Fox. *The Life of John Locke*. London: Henry S. King, 1876.

Boyle, Robert. *The Works of Robert Boyle*, ed. Michael Hunter and Edward B. Davis. London: Pickering and Chatto, 1999–2000.

Cranston, Maurice. *John Locke*. London: Longmans, 1957.

Descartes, René. *Opera Philosophica*. Amsterdam: Louis and Daniel Elzevir, 1654–6.

Descartes, René. *Opera Philosophica*. Amsterdam: Jan Janszoon, 1656–8.

Descartes, René. *Epistolae*. Amsterdam: Daniel Elzevir, 1668.

Evelyn, John. *The Diary of John Evelyn*, ed. E. S. de Beer. Oxford: Clarendon Press, 1955.

Glanvill, Joseph. *The Vanity of Dogmatizing*. London: Henry Eversden, 1661.

Guibert, A.-J. *Bibliographie des oeuvres de René Descartes publiées au XVIIe siècle*. Paris: Centre National de la Recherche Scientifique, 1976.

King, Lord. *The Life of John Locke*. London: Henry Colburn and Richard Bentley, 1829.

Le Clerc, Jean. 'Eloge de feu Mr Locke', *Bibliothèque choisie*, 6 (1705), 342–411.

Le Clerc, Jean. *Epistolario*, ed. Maria Grazia Sina and Mario Sina. Florence: Leo S. Olschki, 1987–97.

Locke, John. 'Méthode nouvelle de dresser des recueuils, communiquée par l'auteur', *Bibliothèque universelle et historique*, 2 (1686), 315–40.

Locke, John. *A Letter to Edward Lord Bishop of Worcester*. London: A. and J. Churchill, 1697.

Locke, John. *Posthumous Works of Mr. John Locke*. London: A. and J. Churchill, 1706.

Locke, John. *The Works of John Locke*, Esq. London: S. Birt and others, 1751.

Lough, John. 'Locke's Reading during his Stay in France (1675–79)', *The Library*, 8 (1953), 229–58.

Milton, J. R. 'The Date and Significance of Two of Locke's Early Manuscripts', *Locke Newsletter*, 19 (1988), 47–89.

Milton, J. R. 'Locke at Oxford', in G. A. J. Rogers, ed., *Locke's Philosophy: Context and Content*. Oxford: Clarendon Press, 1994.

Milton, J. R. 'John Locke's Medical Notebooks', *Locke Newsletter*, 28 (1997), 135–56.

Milton, J. R. 'The Dating of "Adversaria 1661"', *Locke Newsletter*, 29 (1998), 105–17.

Rogers, G. A. J. *Locke's Enlightenment*. Hildesheim: Georg Olms, 1998.

Ware, Charlotte S. 'The Influence of Descartes on John Locke: A Bibliographical Study', *Revue internationale de philosophie*, 4 (1950), 210–30.

Woolhouse, Roger. 'Lady Masham's Account of Locke', *Locke Studies*, 3 (2003), 167–93.

Woolhouse, Roger. *Locke: A Biography*. Cambridge: Cambridge University Press, 2007.

2

Locke and Cartesian Cosmology

Peter R. Anstey

The Cartesian vortex theory was, arguably, the most widely discussed feature of Descartes's philosophy in late seventeenth-century England.[1] It was the centrepiece of his natural philosophy and the subject of Parts 3 and 4 of his *Principia philosophiae* (1644). It spawned not only intense philosophical and theological debate, but also a whole new genre of cosmogonical writing that flourished in France and England in the latter half of the seventeenth and early eighteenth centuries. Not surprisingly, therefore, the Cartesian vortex theory was of great interest to John Locke. Yet, the evidence for Locke's interactions with and attitude towards this theory has never been assembled. It is this that I aim to do in this chapter.

After providing the briefest of sketches of the Cartesian theory of vortices in Section 1, I provide a preliminary survey of Locke's increasingly critical responses to it in Section 2. Taken in isolation, these comments might suggest that Locke firmly rejected the vortex theory by the late 1680s; however, I argue in Section 3 that a more fine-grained analysis indicates that Locke is typical of many in England who acquiesced in a generic vortical explanation of the system of the planets until after the mid-1690s. In Sections 4 and 5 I survey Locke's broader interest in French and English cosmogonical writings in the 1690s respectively, and finish with some concluding remarks.

1. The Cartesian Theory of Vortices

Descartes provided a comprehensive mechanical model for the origins, structure, and behaviour of the entire material universe. As such it was both a cosmogony, that is, an account of the world's origins, and a cosmology, an account of the structure of the world. The theory is adumbrated in his early *Le Monde* and is developed in full in the *Principia philosophiae* of 1644.[2] The central thesis is that each star lies at the centre of a

[1] For a good general treatment of the English reception of the Cartesian vortex theory, which surprisingly omits any reference to Locke, see Harrison, 'The Influence of Cartesian Cosmology in England'.

[2] For discussion of the theory in *Le Monde*, see Schuster, *Descartes-Agonistes: Physico-mathematics, Method & Corpuscular-Mechanism 1618–33*.

huge fluid vortex,[3] a vortex that rotates in the same direction and along the same axis as its central star. The universe, or world, is an indefinitely large assemblage of these vortices which are largely discrete but which are packed together such that the shape of each vortex is slightly distorted by the pressure of its immediate neighbours. (Imagine a room packed full of inflated balloons each of which represents a vortex.) Planets are large spherical objects which float in a vortex and are carried along by the circumambient fluid matter. Comets are planet-like objects that do not circulate within one vortex but find their passage through the system of vortices, moving in and out and progressing from one vortex to another. Should a comet get sucked into a vortex it would become a planet.

Now Descartes has a clever explanation of how vortices and planets are formed, and he uses the theory to explain most of the known celestial phenomena of his day and many terrestrial phenomena as well. For example, the Cartesian vortex theory provides an explanation of the different orbital speeds of the planets, of the motion of the tides, of magnetism, and, most importantly, of gravity. Our Earth is a planet in a vortex rotating around the Sun. The Earth was once a star at the centre of its own vortex around which the Moon rotated as a planet, but at some point the sun that is now Earth 'degenerated' through the formation of a hard surface crust and it was absorbed with its planet (the Moon) into the Sun's vortex where it is situated today.

A theory such as this was badly needed. The Aristotelian–Ptolemaic cosmology of concentric spheres with the Earth at its centre had been seriously undermined, and the Copernican and Tychonian alternatives were, in Descartes's day, coming to be regarded as explanatorily superior. However, neither theory came with an explanation of the motions of the heavens. It was this that Descartes's mechanical model provided. It was an intuitive and enormously appealing natural philosophical theory. Yet cosmogony and cosmology were not the sole province of the natural philosophers, for the theologians and writers of literature also had a stake in these matters. The Book of Genesis remained the standard creation narrative and history of the early Earth of the day, and by the time that the *Principia philosophiae* appeared there was already a fledgling genre of what we might call extra-terrestrial speculation, such as John Wilkins's *The Discovery of a World in the Moone* (1638).[4] What Descartes's theory did was to unify these three fields of natural philosophy, theology, and literature by providing a fecund template for further elaboration and speculation about the origins, structure, and contents of the universe. And so a new genre of Cartesian-inspired cosmology was born. Cartesian-style cosmologies began to appear in the 1650s, such as Beaufort's *Cosmopoea divina* of 1656, and were still being produced well into the eighteenth century. Thus, any appraisal of Locke's view of the Cartesian vortex theory must take into account not only

[3] The best general treatment of the Cartesian vortex theory remains Aiton, *The Vortex Theory of Planetary Motions*.

[4] For the intersection of fiction and natural philosophy in seventeenth-century cosmological writings, see Aït-Touati, *Fictions of the Cosmos: Science and Literature in the Seventeenth Century*.

Descartes's own theory, but also the many other related cosmologies and cosmogonies that Locke read and even commented upon.

Before proceeding to Locke's reactions to the Cartesian vortex theory, however, it will be useful to lay out the range of criticisms that the theory was subject to as the seventeenth century unfolded, and to gauge, at least in a general way, the reception of the theory in England. There were four main critical responses to the Cartesian theory. First, there was widespread concern that the cosmogony that Descartes presented in the *Principia philosophiae* could not be reconciled to the Genesis account of Creation. Interestingly, there is no explicit evidence that this was a reason for the inclusion of all of Descartes's works on the Index in 1663,[5] though no doubt for some it was certainly in the background. Needless to say, defenders were soon to be found, not least the Dutchman Johannes Amerpoel, who in a parallel treatment sought to demonstrate the compatibility of the two cosmogonies in his *Cartesius Mosaizans* of 1669. Another who took on this task was Géraud de Cordemoy in his *Copie d'une lettre écrite à un sçavant religieux* of 1668.

Second, and most important for our enquiry, is criticism that Descartes's theory was inconsistent with Kepler's second and third laws of planetary motion. We will return to this in Section 2. Third, some claimed that there were weaknesses in the theory's explanations of particular phenomena such as the motion of the tides and the nature of gravity.[6] And fourth, it was claimed that the theory was incompatible with the existence of the void, and that there were good reasons for thinking that the inter-planetary regions were in fact empty space.

During the first three decades after it appeared, none of these criticisms was so obvious or pressing that the theory did not find widespread acceptance in England in the latter half of the seventeenth century. Henry More was the earliest and most enthusiastic advocate of the vortex theory, providing a detailed account of it in 1647.[7] Robert Boyle provides a useful barometer of the uptake of the theory amongst English natural philosophers, not least because of his close friendship with Locke. He spoke of 'this vortex of ours' in his *Excellency of Theology* of 1674 and 'the present System of our *Vortex*' as late as 1686 in his *Notion of Nature*.[8] By 1685 so widespread was the acceptance of the vortex theory that Richard Blome could entitle a section of his *The Gentleman's Recreation* as 'A brief account of Des Cartes's Philosophy; being the New and General received philosophy'.[9] So what did Locke make of all of this?

[5] See Armogathe and Carraud, 'The First Condemnation of Descartes's *Œuvres*: Some Unpublished Documents from the Vatican Archives'. However, Descartes's commitment to Copernicanism was regarded as a point of inconsistency with Catholic doctrine, see ibid., p. 76.

[6] For criticism of the Cartesian theory of gravity, see, for example, Keill, *Introductio ad veram physicam*, sigs b1^{r}–b3^{r}.

[7] More, *Philosophical Poems*, pp. 396–400.

[8] *Works of Robert Boyle*, 8, p. 21 and 10, p. 508. *Excellency of Theology* was composed around 1665. See *Works of Robert Boyle*, 8, p. 5. See also Cumberland, *De legibus naturae*, pp. 41 and 320.

[9] Blome, *A Gentlemans Recreation*, p. 43. The discussion of the vortex theory is on p. 44. The stationer Blome later undertook to publish Antoine Le Grand's *An Entire Body of Philosophy, According to the Principles of the Famous Renate Des Cartes*.

2. Locke and Descartes's Vortex Theory

The earliest reference to the Cartesian vortex theory that I have found in Locke's manuscripts is a Latin entry in his journal made on 1/11 August 1677 while Locke was in France. He asks about the origins of Descartes's theory: 'Videre est Cartesii vortices apud Jordanum Brunum de immenso...' [See that the vortices of Descartes are from Giodano Bruno's *De immenso*].[10] The earliest journal entry that indicates his attitude towards the Cartesian vortex theory was made on 1/11 February 1678/9:

Mr Toinard shewd me a new systeme of our Tourbillion wherein the center of the sun described a circle of the turbillion in which it made its periodical circuit in 6 months & Mercury moved about the sun as the moon doth about the earth.[11]

Written toward the very end of Locke's prolonged sojourn in France, this entry suggests pretty clearly that Locke acquiesced in the vortex theory at this time. That the language of vortices was now embedded in Locke's consciousness is seen in the following extract of a letter to Toinard just months after viewing Toinard's 'new systeme':

On leaving you I became restless and I believe that our island is a vortex carried with more agitation than the rest of the Earth. I do not know if you think that I am a little too Cartesian in this, but I feel that I have a little too much of the 'machine', and that in this great agitation in which I have found myself since my return, my mind also has not had that tranquility necessary for a conversation as philosophical as yours.[12]

It is clear, however, that Locke did not identify himself as a Cartesian at this point. A year later Locke was again distancing himself from the Cartesians. He writes to Toinard from London:

Last Tuesday hailstones of enormous size fell all over the city here. I myself measured one lump of ice (for that is what they really were) which had a circumference of 420 grys; it was rounded in shape and slightly flattened on both sides, so that it was not perfectly spherical. I hear that others were measured by various people and found to have twice as great a circumference; but the middling specimen that I handled myself sufficiently astonished me, and I should be glad to know from your philosophers up to what weight solid bodies of such bulk can be suspended in the air. I doubt whether the Cartesians can have any contrivances to help in this matter, and whether the Occult Qualities of the Peripatetics may not break down under such a load.[13]

[10] Bodleian Library (hereafter 'Bodl.') MS Locke f. 2, p. 229; see Lough, *Locke's Travels in France*, p. 162.

[11] British Library Add. MS 15642, pp. 8–9.

[12] Locke to Nicolas Toinard, 25 May 1679 (old style), LC 2, p. 25: 'En vous quittant j'ay perdue toute sorte de repos, et je crois que nostre isle est un tourbillion emporté avec plus d'agitation que tout le rest de la terre, Je ne scay si vous me croirez en cela un peu trop Cartesien mais je sen bien que j ay un peu trop de la machine, et que dans la grand agitation dans laquelle je me suis trouvé depuis mon retour mon esprit aussi n'a pas eu la tranquillité necessaire a une conversation si philosophique que la vostre.'

[13] Locke to Nicolas Toinard, 20/30 May 1680, LC 2, pp. 175–6.

Not only does Locke not align himself with the Cartesians, he doubts that they could explain the phenomenon of the giant hailstones on their own principles.

The next, and indeed far and away the most important, reference to the vortex theory in Locke's writings is found in his review of Newton's *Principia* which appeared in volume 8 of Jean LeClerc's *Bibliothèque universelle et historique* of 1688.[14] While the review is largely a summary contents list of Books One and Two, along with tables of cosmographical and geodesic data derived from Book Three, it does focus on Newton's arguments against the Cartesian vortex theory in the final scholium of Book Two, Section Nine. Locke reproduces two of Newton's arguments against the theory. The first argument maintains that the Cartesian vortex theory contradicts Kepler's area law: it requires planets to move faster when further from the Sun and slower when closer to it, while Kepler's area law requires the opposite. Then Newton mounts an empirical argument against the theory, claiming that Descartes's theory requires the Earth to speed up as it approaches the perihelion of Mars' orbit because the bands of the vortex are more compressed at that point. However, the observational evidence at the time Newton was writing showed that the opposite is the case. I have analysed the details of Newton's arguments in this scholium elsewhere, so they need not detain us here.[15] The point to glean from all of this is that Locke carefully reproduces them, as well as Newton's diagram at the end of the second book of the *Principia*. Locke opens this section of his review by saying:

> At the end of this section, the author proves that the planets are not carried by corporeal vortices.[16]

He ends the section by claiming:

> The author concludes from this that the hypothesis of vortices does nothing to explain the movements of the celestial bodies and he gives another one in the following book.[17]

It is clear from the contents of the review that Locke followed Newton's reasoning and, from the top and tail of the discussion, that Locke was convinced by Newton's arguments. It does appear, then, that by 1688 Locke was convinced that Descartes's vortex theory had been disproved by Newton, even if at this stage he did not fully appreciate his compatriot's alternative theory.

This conclusion is reinforced a decade later in the Stillingfleet correspondence in which Locke refers the reader back to this very section in the *Principia*. In his *Mr Locke's Reply to the Right Reverend the Lord Bishop of Worcester's Answer to his*

[14] Locke, Review of *Philosophiae Naturalis Principia Mathematica.*

[15] See Anstey, *John Locke and Natural Philosophy*, pp. 93–7.

[16] Locke, Review of *Philosophiae Naturalis Principia Mathematica* p. 440: 'Sur la fin de cette Section, l'Auteur prouve que les Planetes ne sont pas emportées par des tourbillons corporels.'

[17] Locke, Review of *Philosophiae Naturalis Principia Mathematica* , pp. 442–3: 'L'Auteur conclut de là que l'hypothese des Tourbillons ne sert de rien à expliquer les mouvemens des corps célestes, & en donne un [*sic*] autre dans le Livre Suivant.'

Second Letter (1699) Locke is replying to Stillingfleet's claim that we should not proceed by mathematical demonstrations in natural philosophy. Locke replies:

Your lordship's two reasons against this way of advancing knowledge upon the sure grounds of mathematical demonstration, are these:
(1.) 'That Des Cartes, a mathematical man, has been guilty of mistakes in his system.' Answ. When mathematical men will build systems upon fancy, and not upon demonstration, they are as liable to mistakes as others. And that Des Cartes was not led into his mistakes by mathematical demonstrations, but for want of them, I think has been demonstrated by* some of those mathematicians who seem to be meant here.
* Mr. Newton Phil. Natur. Princip. Mathemat. l. 2. §9.[18]

Later in the same work Locke refers to this again, this time in the context of the idea of substance:

For that which I am charged with is the discarding substance. But the discarding substance is not the discarding the notion of substance. Mr. Newton has discarded Des Cartes's vortices, *i.e.* laid down principles from which he proves there is no such thing; but he has not thereby discarded the notion or idea of those vortices, for that he had when he confuted their being, and every one who now reads and understands him will have.[19]

This is pretty decisive evidence of Locke's mature view of the matter, and its origins clearly lie with the *Principia* review of 1688.

At first sight, then, the picture appears to be one of a gradual move on Locke's part from tentative acquiescence in the late 1670s, to scepticism, and then to outright denial by 1688. It looks as if Locke came to reject the Cartesian vortex theory around the time he was completing the *Essay* in the Netherlands, and that in this he was ahead of most of his compatriots. This would sit well with the view that as a proponent of experimental philosophy, Locke was opposed to speculative systems of natural philosophy, and Newton had given him cause to reject the very system that had come to predominate in England, the Cartesian vortex theory. Such a judgement, however, would be a little too swift.

3. The Reception of the Vortex Theory in England

Hints that Locke was not implacably opposed to cosmological theories like that of Descartes are not hard to find. Locke says in a letter to William Molyneux of 2 July 1695 (old style), that is, well after the *Principia* review, speaking of the young Francis Masham:

the child above mentioned, but nine years old in June last, has learn'd to read and write very well; is now reading Quintus Curtius with his mother, understands geography and chronology

[18] Locke, *Second Reply*, LW 4, pp. 426–7.

[19] Locke, *Second Reply*, LW 4, p. 451.

very well, and the Copernican system of our vortex; is able to multiply well, and divide a little; and all this without ever having had one blow for his book.[20]

This is not the statement of one facing off against the vortex theory, even if, as he surely is, Locke is reverting to common parlance. Molyneux does the very same thing himself in his *Dioptrica Nova* of 1692. In this book, which is imbued with the experimental philosophy—and ironically, in the very paragraph containing his discussion of Newton's demonstration that the planets obey Kepler's Third Law—he says the following:

before I leave *Saturn* and *Jupiter*, I cannot but take notice of one admirable Property, for the Knowledge whereof, we are beholden to the Telescope; and that is, the wonderful Agreement which is found in all the several Systems of our Vortex.[21]

It is also striking that when Richard Bentley, the first Boyle lecturer, corresponded with Newton in 1692/3, conceiving of the universe as a system of vortices, Newton's replies contain no denunciation of the vortical framework in which Bentley's cosmology is set.[22] Newton may have disproved Descartes's vortex theory, but vortices still lingered in everyday language, even amongst those who were in the know.[23]

What this suggests is that, while *Descartes*'s theory of vortices had been shown to be merely unfounded speculation, there remained a widespread acceptance in England that the Sun and planets were situated in some kind of vortex, even if it did not operate according to Cartesian principles. It is well known that on the Continent both Leibniz and Christiaan Huygens, by the early 1690s, had developed their own non-Cartesian vortex theories. Leibniz's theory appeared in his 'Tentamen de motuum coelestium causis' in the *Acta eruditorum* in 1689.[24] Huygens's theory appeared in the posthumous *Cosmotheoros* in 1698, but was developed in the early 1690s. And while these theories were not discussed in England until later in the decade, the evidence does point to an ongoing English belief in a generic conception of a vortex that parallels that on the Continent. Indeed, there is very little evidence that in the early to mid-1690s Newton's theory was viewed as having deposed the vortex theory. Instead the gaze of many was fixed on the more accessible Cartesian-inspired cosmogonies of Thomas Burnet, John Woodward, and William Whiston (to be discussed in Section 5), rather than on the technically demanding work of Newton.

It was only around 1697 that the issues crystallized such that it became clear that one needed to side either with vortices or with Newtonian gravity, because the two explanations of the planetary system were incompatible. Clear evidence of this is

[20] LC 5, p. 406, underlining added. [21] Molyneux, *Dioptrica Nova*, p. 272.

[22] See, for example, Richard Bentley to Newton, 18 February 1692 (old style) and Newton to Bentley, 25 February 1692 (old style), *Correspondence of Isaac Newton*, 3, pp. 246–56.

[23] For Locke's use of the term 'vortex' in *An Essay concerning Human Understanding* (1690), see E IV. viii. 3.

[24] Leibniz, 'Tentamen de motuum coelestium causis' (LL 8). See the discussion of Huygens's theory and the two versions of Leibniz's theory in Koyré, *Newtonian Studies*, pp. 115–38 and of Leibniz's theory in Aiton, *The Vortex Theory of Planetary Motions*, pp. 125–51.

found in John Edwards's *Brief Remarks upon Mr. Whiston's* New Theory of the Earth (1697). Edwards says,

There never was a greater dissension among Natural Philosophers than there is about this one thing, as appears from comparing the Writings of *Des Cartes* (who was the First Substantial Author of *Mechanism*) Dr. *More*, Dr. *Burnet*, Mr. *Newton*, Dr. *Woodward*, Mr. *Whiston*, and other Philosophical Men of this age, who have built their respective and different Hypotheses and Theories on their different notions of *Mechanick* Principles and Agents. They *all* pretend to proceed upon *Mechanical* Laws: their Solutions are founded on the Natural Tendency of Matter: and yet we see how wide their Notions are from one another, and how their Hypotheses are contradictory to each other.[25]

It was abundantly clear to Edwards that one had to choose between Descartes and Henry More on the one hand, and Newton on the other:

The short of all is, if you strike in with *Cartes* and *More*, you must renounce the *Lucasian* Professor: if you make choice of this last, you must cashier the other two.[26]

And again,

if *Vortices* cannot do it, *Gravity* must.[27]

It is unlikely that Locke saw the choice so clearly until the late 1690s, for even though he believed by 1688 that Newton had undermined Descartes's vortex theory, he continued to purchase and to read works inspired by the theory. Let us examine his interest in French works first.

4. Locke and the French Cartesian Cosmologies

Locke received a package of French books from Pembroke on 11 October 1689.[28] Among them was a first edition of Fontenelle's *Entretiens sur la pluralité des mondes* (LL 1151[29]), which had first appeared in 1686. This work was enormously popular in England, being translated as early as 1688 by both Aphra Behn and John Glanvill.[30] Locke had almost certainly read the work because William Grigg wrote to him in 1694 saying:

I have spent some hours in translating the Entretiens sur la pluralité des mondes, a Diversion I remember you approv'd of as usefull and agreable, but t'other Day I accidentally met with a printed Translation of it which has put a stop to mine.[31]

[25] Edwards, *Brief Remarks upon Mr. Whiston's* New Theory of the Earth, p. 37. This book was among the eight books by Edwards that Locke owned (LL 1029).

[26] Ibid., p. 45. [27] Ibid., p. 40. [28] See LC 3, p. 789.

[29] All catalogue numbers of works in Locke's library derive from Harrison and Laslett, *The Library of John Locke*, and are prefixed by 'LL'.

[30] Fontenelle, *A Discovery of New Worlds*, trans. Aphra Behn; *A Plurality of Worlds*, trans. John Glanvill.

[31] William Grigg to Locke, 16 June 1694 (old style), LC 5, p. 77.

That he approved of Grigg's translation efforts is a mark of Locke's wider interest in Cartesian-inspired cosmologies. For, not only did he own Fontenelle, but also the 1670 English translation of Cordemoy's defence of Descartes's cosmogony (LL 845[a]), the 1675 edition of Rohault's *Traité de physique* (LL 2497), and all three volumes of Claude Perrault's *Essais de Physique* (LL 2259) which Toinard sent him in May 1680. The third of these volumes makes extensive use of the Cartesian vortex theory.[32] Another lesser-known work which is not listed in Harrison and Laslett, but which Locke records lending to Boyle on 13/23 December 1679, is P. Chérubin's *Effets de la force de la contiguité des corps*, Paris 1679.[33] This plenist work also presupposes the Cartesian vortex theory and would have been of great interest to Boyle as it discusses his own work and Linus' funicular hypothesis.[34]

Yet, Locke was also aware of anti-Cartesian works in French. Pieter Guenellon wrote to him in August 1693 commenting:

> Mr Verrin has just translated *The Voyage of the World of Descartes* that you will no doubt have seen in French in octavo and which contains some fine objections against the system of Descartes. The style is very playful and it is said that the author is a Jesuit, but I am telling you what you no doubt know better than me.[35]

He is, of course, referring to Gabriel Daniel's satirical *Voyage du monde de Descartes* which Locke owned (LL 913, 1691 edition). It is clear then that from the late 1670s Locke was keeping abreast of some of the main French Cartesian-inspired cosmologies and some of critical literature as well.[36] Perhaps it was this that piqued his interest in the English cosmogonies, for from the mid-1680s Locke developed a sustained interest in these writings and even had a modest impact on the genre itself. Let us turn then to England.

5. Locke and English Cosmogonies

Not all of the Cartesian cosmogonies were French.[37] For example, Locke owned Thomas Robinson's Cartesian *New Observations on the Natural History of this World* (1696). But our interest lies not with specifically Cartesian views but with cosmogonies that were inspired by Descartes. These cosmogonies focused on the history of the Earth rather than the origins of the planetary system.

[32] Toinard to Locke, 12/22 May 1680, LC 2, p. 166.

[33] British Library Add. MS 15642, p. 185.

[34] Locke met the Capuchin friar Chérubin d'Orléans on 5/15 February 1678/9, see British Library Add. MS 15642, p. 31.

[35] Pieter Guenellon to Locke, 2/12 August 1693, LC 4, p. 708.

[36] Locke's interest in fictional cosmological literature extends back to *c.*1659, for at that time Locke owned Cyrano de Bergerac's *Selenarchia, or The Government of the World in the Moon* and Wilkins's *The Discovery of a World in the Moone*. See Bodl. MS Locke f. 14, pp. 8–9, Bodl. MS Locke d. 10, p. 259, and Strachey to Locke, 22 November 1660 (old style), LC 1, pp. 159–60. J. R. Milton alerted me to the manuscript references.

[37] For further discussion of Burnet, Woodward, and Whiston, see Poole, *The World Makers*, ch. 5.

The first English cosmogony that Locke read and commented upon was Burnet's *Theory of the Earth*. Originally published in Latin in 1681 as *Telluris theoria sacra*, Locke probably read the 1684 English translation of the first two volumes (LL 534) and commented on it in detail in a letter to James Tyrrell of 14/24 February 1687. He was dissatisfied with Burnet's account of the Deluge and claimed 'I could give you an hypothesis would explain the deluge without half the difficulties.'[38] And that is exactly what he did in *Some Thoughts Concerning Education* in which he speculated that to cause the Deluge God moved the Earth's centre of gravity for a time, moving it around to create flooding all over the Earth.[39]

John Woodward disagreed with Locke's hypothesis and drafted a tract attacking it. We learn this from a letter from Alexander Beresford to Locke who reports:

> Dr Woodward oppose's your Hypothesis about the Deluge (wherein by the way I hear you are not single) only en passant in his Essay, but in another Treatise (merely on Gravity) already finish't, though not printed, I hear he professedly undertake's it, and pretend's to Demonstrate that a Change of the Gravity-Center could produce, at most, but a Partiall deluge.[40]

Woodward does indeed allude critically to Locke's theory in his *An Essay towards a Natural History of the Earth* which came out in the same year.[41] But Locke found support for his view in both the first Boyle Lecturer, Richard Bentley, and William Whiston, as we hear from William Molyneux:

> Mentioning these books minds me to Intimate to you, that these Ingenious Authors agree also exactly with you in a Passage you have in your *Thoughts of Education* pag. 337. 3d Edit. Sec. 192. That the Phænomenon of Gravitation cannot be Accounted for by Meer Matter and Motion, but seems an Immediate Law of the Divine Will so ordering it. And you conclude that Section thus—*Reserving to a fitter Oppertunity a fuller Explication of this Hypothesis, and the Application of it to all the Parts of the Deluge, and any Difficultys can be supposed in the History of the flood.* This seems to Imply that you have some thoughts of Writing on that subject; it would be a mighty satisfaction to me to know from you the certainty thereof. I should be very Glad also to hear what the Opinion of the Ingenious is concerning Mr Whistons Book.[42]

Molyneux was referring to William Whiston's *A New Theory of the Earth* (1696), which posited that the Deluge and other events in the Earth's history were caused by interactions with comets, all written in a style mimicking Newton's *Principia*. Locke soon replied to Molyneux giving him his opinion of Whiston's book.[43] We should be aware, however, that there is a degree of mutual backslapping going on here, for Whiston praises Locke in his book.[44] Nevertheless, in the midst of all of this cosmogonical

[38] Locke to James Tyrrell, LC 3, p. 140.

[39] Locke, *Some Thoughts Concerning Education*, pp. 246–7.

[40] Alexander Beresford to Locke, 24 March 1695 (old style), LC 5, p. 299.

[41] Woodward, *An Essay towards a Natural History of the Earth*, p. 43.

[42] William Molyneux to Locke, 26 September 1696 (old style), LC 5, p. 702.

[43] Locke to William Molyneux, 22 February 1697 (old style), LC 6, p. 6. For further discussion of Locke and Whiston, see my *John Locke and Natural Philosophy*, pp. 101–3.

[44] Whiston, *A New Theory of the Earth*, Book 2, p. 128.

speculation going on in England, the tide was moving against the Cartesian vortex theory and the speculative cosmologies that it had inspired.

The modes of critical evaluation of the various theories varied, but in every case the Cartesian theory fared badly. One approach was to deploy the rhetoric of the new experimental philosophy and to brand the various theories as speculative hypotheses. A pertinent example of this approach is Robert Saint Clair's *The Abyssinian Philosophy Confuted* (1697). Saint Clair had translated Bernardino Ramazzini's treatise on the Modena springs into English, and presented it as evidence against Burnet's theory of the antediluvian Earth. He also included a polemical essay of his own against Burnet's theory and some other experimental evidence, citing along the way the authority and works of Robert Boyle. This work then bears all the hallmarks of the new experimental philosophy: it presents observational evidence in the form of a natural history, as well as other experimental observations, against a speculative hypothesis, and it is replete with the rhetoric of experimental philosophy:

> I shall not care for the displeasure of these men of *Ephesus* [Burnet and others], whose trade it is to make Shrines to this their *Diana* of Hypothetical Philosophy, I mean who in their Closets make Systems of the World, prescribe Laws to Nature, without ever consulting her by Observation and Experience, who (to use the Noble Lord *Verulams* words) like the Spider ... spin a curious Cob-web out of their Brains ...[45]

Of course, Newton was later to accuse Descartes and Leibniz of being hypothetical philosophers as opposed to experimental philosophers.[46] And all of this would have struck a chord with Locke who favoured experimental philosophy, particularly that which was practised according to the method of natural history, and opposed speculative hypotheses.[47] It is all the more interesting, therefore, to find in Locke's journal that Robert Saint Clair sent him a copy of his book in January 1697.[48] Saint Clair, who lived in Robert Boyle's household at the time of the latter's death and was named in and a signatory to his will,[49] probably knew of Locke's reservations about Burnet's theory and surmised that he would be interested in further critical reflections upon it. Or perhaps he knew that Locke would be interested in the chemical contents of the material appended to the work, for it was almost certainly he who had promised twice to procure for Locke some of Boyle's 'red earth' soon after Boyle's death.[50]

[45] Saint Clair, *The Abyssinian Philosophy Confuted*, sig. a4. There is some overlap in content with 'An Extract of a Letter from Dr. Robert St. Clair to Dr. Rob. Hooke, F. R. S.'.

[46] Newton to Roger Cotes, March 1713, *Correspondence of Isaac Newton*, 5, pp. 398–9.

[47] See Anstey, *John Locke and Natural Philosophy*, *passim*.

[48] See Locke's journal, Bodl. MS Locke f. 10, p. 342: 'The Abyssinian philosophie confuted, from the author 1697.'

[49] Saint Clair speaks of a form of Phosphorus which he saw 'when living with the honourable and never to be forgotten Mr. *Boyle*', 'Of Phosphorus' in 'To the Reader' appended to *The Abyssinian Philosophy Confuted*. See Maddison, *The Life of the Honourable Robert Boyle*, Appendix 1 and LC 4, p. 388.

[50] See Robert Pawling to Locke, 16 January 1692, LC 4, pp. 364–5 and Pawling to Locke 18 February 1692 (old style), LC 4, p. 388. For Boyle's red earth and its relation to the Sophic Mercury, see Principe, *The Aspiring Adept: Robert Boyle and his Alchemical Quest*, p. 177 and Anstey, 'Newton and Locke'. For Jean Le

Yet, for all of this, it seems not to have been the methodological critiques that proved decisive against the vortex theory. By 1698 the Newtonian John Keill had joined the fray and he skilfully and effectively revived Newton's critique of the vortex theory itself. Keill attacked Burnet, Whiston, and Descartes in his *An Examination of Dr Burnet's* Theory of the Earth. *Together with some Remarks on Mr Whiston's* New Theory of the Earth of 1698. Here is what he says about Descartes:

So far was *Des Cartes* from Marrying Physicks with Geometry, that it was his great fault that he made no use at all of Geometry in Philosophy.... [Descartes] was so far from applying Geometry and observations to natural Philosophy, that his whole System is but one continued blunder upon the account of his negligence in that point. This I can easily prove by shewing that his Theory of the *Vortices*, upon which his Systeme is grounded, is absolutely false. The great Philosopher of this age, the most Ingenious and Incomparable Mr. *Newton* by his great and deep skill in Geometry, has shewed that the periodical times of all Bodies which swim in a *Vortex*, must be directly as the squares of their distances from the center of the Vortex. But it is evident from observations, that the Planets in turning round the Sun, observe quite another sort of a law than this, for the squares of their Periodical times, are always as the cubes of their distances, and therefore since they do not observe that law, which of necessity they must, if they swim in a *Vortex*, it is a demonstration that there are no vortices, in which the Planets are carried round the Sun.

Besides if the earth were carried in a *vortex*, it must necessarily move faster, when it is in the beginning of *Virgo*, where the fluid is in a narrow space, (and by consequence moves so much the swifter,) than it would do when it is in the beginning of *Pisces*, and that in the proportion of three to two, which is directly against experience, and observation.

It is impossible therefore upon this, and a great many other accounts, which Mr. *Newton* has shew'd in his principles, that the earth and the other planets can move in a *Vortex*. So that the notion of a *Vortex* being ruined, the whole *Cartesian* system must of necessity fall to the ground; and that world, whose origination he pretended to have deduced from Mechanical principles, must be a wild chimera of his own imagination.[51]

Keill directs his argument against the vortex theory in general and stresses its consequences for Descartes: 'the notion of a *Vortex* being ruined, the whole *Cartesian* system must of necessity fall to the ground'. Newton had shown that vortex theories were inconsistent both with Kepler's third law and with the empirical evidence of the speed of the planet in Virgo. Of equal interest, however, is the fact that Keill's claim about Descartes's failure to apply geometry to natural philosophy is exactly the same as the point that Locke makes against Descartes in the Stillingfleet correspondence:

When mathematical men will build systems upon fancy, and not upon demonstration, they are as liable to mistakes as others. And that Des Cartes was not led into his mistakes by mathematical demonstrations, but for want of them, I think has been demonstrated by some of those mathematicians who seem to be meant here.[52]

Clerc's comments to Locke on Whiston's and Burnet's writings, see Le Clerc to Locke, 7/17 November 1697, LC 6, pp. 253–4.

[51] Keill, *An Examination of Dr Burnet's* Theory of the Earth, pp. 15–17.

[52] Locke, *Second Reply*, LW 4, p. 427.

It is tempting to conclude that Locke, and perhaps even Stillingfleet, had been reading Keill on this very point. One who certainly read Keill was John Sergeant, Locke's adversary in *Method to Science* and *Solid Philosophy Asserted*. In his *Non ultra: or a Letter to a Learned Cartesian*, Sergeant cites this very passage as decisive evidence against the Cartesian vortex theory.[53] Whiston replied to Keill in his *A Vindication of the* New Theory of the Earth (1699), and Keill wrote a rejoinder in *An Examination of the Reflections on the* Theory of the Earth (1699). And it is almost certain that Locke kept abreast of all of these developments because he owned all of these books:

1. Thomas Burnet, *Theory of the Earth*, Vol. 1, London, 1684 (LL 534)
2. John Woodward, *An Essay towards a Natural History of the Earth*, London, 1695 (LL 3179)
3. William Whiston, *A New Theory of the Earth*, London, 1696 (LL 3142)
4. Thomas Robinson, *New Observations on the Natural History of this World of Matter*, London, 1696 (LL 2490) (Cartesian)
5. Robert Saint Clair, *The Abyssinian Philosophy Confuted*, London, 1697 (LL 2528)
6. John Edwards, *Brief Remarks upon Mr Whiston's* New Theory of the Earth, London, 1697 (LL 1029)
7. William Whiston, *A Vindication of the* New Theory of the Earth, London, 1698 (LL 3143)
8. John Keill, *An Examination of Dr Burnet's* Theory of the Earth. *Together with some Remarks on Mr Whiston's* New Theory of the Earth, Oxford, 1698 (LL 1613)
9. John Keill, *An examination of the* Reflections on The Theory of the Earth, Oxford, 1699 (LL 1612)
10. John Sergeant, *Non ultra: or a Letter to a Learned Cartesian*, London, 1698 (LL 2629)
11. Thomas Burnet, *Reflections upon* The Theory of the Earth, London, 1699 (LL 535).

6. Conclusion

It is clear, therefore, that while by 1688 Locke had come to regard Descartes's speculative vortex theory as having been disproved by Newton, at the same time he maintained an interest in Cartesian-style and Cartesian-inspired cosmogonical writings in both English and French well into the 1690s. It is also most likely that he continued to believe that a vortical explanation of planetary motions was plausible. In fact, he even

[53] John Sergeant, *Non ultra: or a Letter to a Learned Cartesian*, pp. 112–17.

went so far as to recommend them as suitable reading for the young. He tells us in *Some Thoughts concerning Education* of 1693 that:

I cannot say, I know any [natural philosophical system] which can be taught a Young Man as a Science, wherein he may be sure to find Truth and Certainty, which is, what all Sciences give an expectation of. I do not hence conclude that none of them are to be read: It is necessary for a Gentleman in this learned Age to look into some of them, to fit himself for Conversation. But whether that of *Des Cartes* be put into his Hands, as that which is most in Fashion; or it be thought fit to give him a short view of that and several others also, I think the Systems of *Natural Philosophy*, that have obtained in this part of the World, are to be read, more to know the *Hypotheses*, and to understand the Terms and Ways of Talking of the several Sects, than with hopes to gain thereby a comprehensive, scientifical, and satisfactory Knowledge of the Works of Nature.[54]

It was not until the correspondence with Stillingfleet, which coincided with the spate of publications from 1697 on the theory, that Locke came to dismiss any form of vortex theory outright.

If there is a Lockean legacy that arises from his involvement in these natural philosophical debates, it is perhaps his deployment of the term 'solar system' in *Some Thoughts Concerning Education*. I had always taken this term to be a Newtonian *replacement* for 'vortex' but I am now inclined to think, on the assumption that it was in use at the time, that it is carefully neutral, applying equally to both theories. Locke says:

[Newton's] Book will deserve to be read, and give no small light and pleasure to those, who willing to understand the Motions, Properties, and Operations of the great Masses of Matter, in this our Solar System, will but carefully mind his Conclusions, which may be depended on as Propositions well proved.[55]

He would have had no inkling that it would be this term 'solar system' that would become the main English descriptor for what in his day was better known as 'our vortex'.[56]

References

Aït-Touati, Frédérique. *Fictions of the Cosmos: Science and Literature in the Seventeenth Century*. Chicago: University of Chicago Press, 2011.

Aiton, E. J. *The Vortex Theory of Planetary Motions*. London: Macdonald, 1972.

Amerpoel, Johannes. *Cartesius Mosaizans*. Leiden, 1669.

Anstey, Peter R. *John Locke and Natural Philosophy*. Oxford: Oxford University Press, 2011.

[54] Locke, *Some Thoughts Concerning Education*, p. 247, underlining added.

[55] Locke, *Some Thoughts Concerning Education*, p. 249. Richard Bentley also used the term in *The Folly & Unreasonableness of Atheism*, published around the same time and owned by Locke (LL 269), though this probably appeared too late to be Locke's source.

[56] I should like to thank Michael Ayers, Martha Bolton, Daniel Garber, Michael Hunter, J. R. Milton, and Lawrence Principe for comments on an earlier version of this chapter.

Anstey, Peter R. 'Newton and Locke', in Eric Schliesser and Christopher Smeenk, eds., *The Oxford Handbook of Newton*. Oxford: Oxford Scholarship Online, 2017.

Armogathe, Jean-Robert and Carraud, Vincent. 'The First Condemnation of Descartes's *Œuvres*: Some Unpublished Documents from the Vatican Archives', *Oxford Studies in Early Modern Philosophy*, 1 (2003), 67–109.

Beaufort, Louis de. *Cosmopoea divina*. Leiden, 1656.

Bentley, Richard. *The Folly & Unreasonableness of Atheism*. London, 1693.

Blome, Richard. *A Gentlemans Recreation*. London, 1685.

Boyle, Robert. *The Works of Robert Boyle*, 14 vols., ed. Michael Hunter and E. B. Davis. London: Pickering and Chatto, 1999–2000.

Burnet, Thomas. *Telluris theoria sacra*. London, 1684.

Burnet, Thomas. *Theory of the Earth*, Vol. 1. London, 1684.

Burnet, Thomas. *Reflections upon* The Theory of the Earth. London, 1699.

Chérubin d'Orléans (religious name of Michel Lasseré). *Effets de la force de la contiguité des corps*. Paris 1679.

Cordemoy, Géraud de. *Copie d'une lettre écrite à un sçavant religieux*. Paris, 1668.

Cumberland, Richard. *De legibus naturae*. London, 1672.

Cyrano de Bergerac, Savinien de. *Selenarchia, or The Government of the World in the Moon*. London, 1659.

Daniel, Gabriel. *Voiage du monde de Descartes*. Paris, 1691.

Descartes, René. *Principia philosophiae*. Amsterdam, 1644.

Descartes, René. *Le Monde de Mr. Descartes*. Paris, 1664.

Edwards, John. *Brief Remarks upon Mr Whiston's* New Theory of the Earth. London, 1697.

Fontenelle, Bernard Le Bovier de. *Entretiens sur la pluralité des mondes*. Paris, 1686.

Fontenelle, Bernard Le Bovier de. *A Discovery of New Worlds*, trans. Aphra Behn. London, 1688.

Fontenelle, Bernard Le Bovier de. *A Plurality of Worlds*, trans. John Glanvill. London, 1688.

Harrison, John and Laslett, Peter. *The Library of John Locke*, 2nd edition. Oxford: Clarendon Press, 1971.

Harrison, Peter. 'The Influence of Cartesian Cosmology in England', in Stephen W. Gaukroger, John A. Schuster, and John Sutton, eds., *Descartes' Natural Philosophy*. London: Routledge, 2000, pp. 168–92.

Huygens, Christiaan. *Cosmotheoros*. The Hague, 1698.

Keill, John. *An Examination of Dr Burnet's Theory of the Earth. Together with some Remarks on Mr Whiston's New Theory of the Earth*. Oxford, 1698.

Keill, John. *An Examination of the Reflections on The Theory of the Earth*. Oxford, 1699.

Keill, John. *Introductio ad veram physicam*. Oxford, 1702.

Koyré, Alexandre. *Newtonian Studies*. Chicago: University of Chicago Press, 1965.

Le Grand, Antoine. *An Entire Body of Philosophy, According to the Principles of the Famous Renate Des Cartes*. London, 1694.

Leibniz, G. W. 'Tentamen de motuum coelestium causis', *Acta Eruditorum* (February 1689), 82–101.

Locke, John. Review of *Philosophiae Naturalis Principia Mathematica. Auctore Is. Newton, Bibliothèque universelle et historique*, tome 8 (1688), 436–50.

Locke, John. *Mr Locke's Reply to the Right Reverend the Lord Bishop of Worcester's Answer to his Second Letter*. London, 1699.

Lough, John. *Locke's Travels in France*. Cambridge: Cambridge University Press, 1953.
Maddison, R. E. W. *The Life of the Honourable Robert Boyle*. London: Taylor and Francis, 1969.
Molyneux, William. *Dioptrica Nova: A Treatise of Dioptricks*. London, 1692.
More, Henry. *Philosophical Poems*. London, 1647.
Newton, Sir Isaac. *The Correspondence of Isaac Newton*, 7 vols., ed. H. W. Turnbull, J. F. Scott, A. R. Hall, and L. Tilling. Cambridge: Cambridge University Press, 1959–77.
Perrault, Claude. *Essais de Physique*, 3 vols. Paris, 1680.
Poole, William. *The World Makers: Scientists of the Restoration and the Search for the Origins of the Earth*. Oxford: Peter Lang, 2010.
Principe, Lawrence. *The Aspiring Adept: Robert Boyle and his Alchemical Quest*. Princeton: Princeton University Press, 1998.
Robinson, Thomas. *New Observations on the Natural History of this World of Matter*. London, 1696.
Rohault, Jacques. *Traité de physique*. Paris, 1675.
Saint Clair, Robert. *The Abyssinian Philosophy Confuted*. London, 1697.
Saint Clair, Robert. 'An Extract of a Letter from Dr. Robert St. Clair to Dr. Rob. Hooke, F. R. S.'. *Philosophical Transactions*, 20 (1698), 378–81.
Schuster, John A. *Descartes-Agonistes: Physico-mathematics, Method & Corpuscular-Mechanism 1618–33*. Dordrecht: Springer, 2013.
Sergeant, John. *The Method to Science*. London, 1696.
Sergeant, John. *Solid Philosophy Asserted*. London, 1697.
Sergeant, John. *Non ultra: or a Letter to a Learned Cartesian*. London, 1698.
Whiston, William. *A New Theory of the Earth*. London, 1696.
Whiston, William. *A Vindication of the New Theory of the Earth*. London, 1698.
Wilkins, John. *The Discovery of a World in the Moone*. London, 1638.
Woodward, John. *An Essay towards a Natural History of the Earth*. London, 1695.

3

The Cartesian Element in Locke's Anti-Cartesian Conception of Body

James Hill

Perhaps the most distinctive claim in Locke's treatment of body in the *Essay concerning Human Understanding* is his disavowal of any knowledge of the essence of matter. This claim implies a rejection of the optimism of Cartesian epistemology that would treat our faculties as capable of knowing the underlying nature of the material realm. It is also a rejection of Descartes's specific geometrical conception of the essence of matter as extension in length, breadth, and depth. Locke holds that the geometrical conception ignores the solidity of matter, a quality that he then uses to distinguish between body and space, enabling him to deny Descartes's arguments in favour of the plenum and to defend the possibility of a vacuum. What is more, Locke thought that we know the solidity of body by way of the senses, not by way of a Cartesian pure intellect, thus treating our best conception of matter as empirically derived. In the light of all these contrasts, it is surely uncontroversial to characterize Locke's treatment of body as 'anti-Cartesian'.

Yet, as I shall argue here, Locke remains deeply indebted to Descartes. His conception of body evolves in the Drafts of the *Essay* in visible engagement with the Cartesian view. He may oppose Descartes in the different ways we have mentioned, but it would be quite wrong to treat Descartes's thought about body as having only negative significance for him. This is partly because the sharp differences between the two thinkers are made possible by background agreement, but it is also because there is a crucial point of doctrinal detail on which Locke takes the Cartesian line. It is this point of agreement that I am calling the 'Cartesian element' in Locke's approach. One result of recognizing the Cartesian element is that it enables us to understand how Locke differed from the Epicurean-atomist position, to which he has been routinely assimilated.[1]

[1] The *locus classicus* for the Epicurean-atomist interpretation of Locke is Leibniz's *New Essays*. A more up-to-date defence of this interpretation is presented in Lennon, *The Battle of the Gods and Giants*, where we are told that 'Leibniz gets the picture of Locke exactly right' (p. 149). The atomist reading is also a theme in Peter Alexander's interpretation of Locke; see in particular his 'Solidity and Elasticity'.

1. Background Agreement

Before we look at the Cartesian element, let us briefly look at the background agreement we have just mentioned. Locke famously remarked to Lady Masham (as she related in a letter written after his death) that:

> The first Books which gave him a relish of Philosophical studys were those of Descartes. He was rejoyced in reading of these because tho' he very often differ'd in Opinion from this writer, he yet found that what he said was very intelligible: from whence he was incourag'd to think That his not haveing understood others, had, possibly, not proceeded altogether from a defect in his Understanding.[2]

Locke found in Descartes's philosophy an ideal of intelligibility. This ideal was most probably made apparent to him when studying Descartes's account of the material world. The vast majority of the notes that Locke made when reading Descartes pertain to questions about the mechanistic view of nature set out in *Principles*, Parts 2–4, and there is little manuscript evidence that he spent time grappling with the details of Descartes's metaphysics in the *Meditations* and Part 1 of the *Principles*.[3] The comments to Lady Masham may therefore be taken as commending, in the most general way, the application of Cartesian clarity and distinctness to our picture of corporeal nature and its workings.

Locke also came to have, in common with Descartes, some commitment to the mechanical view of the natural world. There has been debate in recent decades about what exactly the mechanical philosophy would have meant for Locke, and the extent to which he was attached to it. But it is generally accepted that, by the time of the *Essay*, Locke favoured mechanistic explanations of natural phenomena over rival explanations, and that this preference was related to their intelligibility.[4] In Draft B he tells us, of causes in general, that 'we can hardly conceive their efficacy to consist in anything but motion',[5] and in the first edition of the *Essay* he opines that it is 'impossible to conceive' that body 'should operate any other way than by Motion'.[6] And while the latter statement was revised somewhat in later editions, its assertion of the special conceivability of mechanical impulse remained. It is the superior intelligibility of mechanism which led Locke to ultimately shun the formal causes of the scholastics and the more exotic powers and virtues advocated by the chymists. The question

[2] Woolhouse, 'Lady Masham's Account of Locke', p. 173.

[3] See Chapter 1 of this book, Milton's 'Locke and Descartes: The Initial Exposure, 1658–1671'. See also Milton, 'Locke at Oxford', where it is concluded, from an examination of Locke's commonplace books, that the passages he copied from Descartes's work were 'overwhelmingly concerned with what we would categorize as natural science rather than philosophy'. In the case of the *Principles*, it is shown that Locke paid particularly close attention to Parts 3 and 4, some attention to Part 2, but took nothing from Part 1 at all.

[4] For the history of Locke's gradual warming to mechanism see Milton, 'Locke, Medicine and the Mechanical Philosophy' and Walmsley, 'The Development of Locke's Mechanism'. One signal point here is that the primary and secondary quality distinction had not featured in Drafts A and B, but only became a part of Locke's thought in Draft C (C. II. vii), which was prepared in the mid-1680s. I would like to thank J. R. Milton for kindly giving me access to his typescript of Draft C.

[5] Draft B, # 138, p. 256. [6] E II. viii. 11, pp. 135–6.

remained, however, as to how far this intelligibility really went. Locke may have found an ideal in Cartesian mechanism, but he was pessimistic about our ability to realize this ideal. Though we may have the idea of 'a perfect *Science* of natural Bodies', Locke ultimately concluded that it was 'lost labour' to pursue it.[7]

2. Hardness

So much, then, for background orientation, but what is the specifically Cartesian element in Locke's conception of body? It is his understanding of hardness, and his distinction between hardness and impenetrability. I shall argue that Locke's view of hardness is not just recognizably derived from Descartes's, but that it has important influence on his general understanding of the material world and on his reasons for maintaining an agnosticism about the essence of matter.

A philosopher's treatment of hardness may seem like a marginal issue to us—a point of interest, perhaps, but not a question of any real significance in the overall philosophical treatment of nature. In the seventeenth century, however, the treatment of hardness was a crucial point of contention dividing the two principal schools of the mechanical philosophy: Cartesianism and Epicureanism. In the Epicurean framework of atoms and the void, hardness was the fundamental property of matter and was the source of the indivisibility of the atoms. Hardness was what made atoms '*atomoi*' or 'uncuttables'. Descartes, on the other hand, proudly and explicitly rejected the indivisibles of Democritus.[8] He treated hardness as an emergent property, not pertaining to bodies per se, but only to configurations of matter that human beings could interact with by the bodily sense of touch. Paying attention to Locke's view of hardness can therefore help us understand his attitude to these two pictures of the material world, and the reasons why he could accept neither.

3. Descartes on Hardness

Descartes considers hardness early in Part 2 of his *Principles*, where he seeks to show that this property—along with weight, colour, and the other phenomenal qualities—does not belong to the essence or nature of body.

For as regards hardness [*duritia*], our sensation tells us no more than that the parts of a hard body resist the motion of our hands when they come into contact with them. If, whenever our hands are moved in a given direction, all the bodies in that area were to move away at the same speed as that of our approaching hands, we should never have any sensation of hardness. And since it is quite unintelligible to suppose that, if bodies did move away in this fashion, they would thereby lose their bodily nature, it follows that this nature cannot consist in hardness.[9]

[7] E IV. iii. 29, p. 560. [8] See, for example, P 4, p. 202: AT 8A, p. 325; CSM 1, p. 287.
[9] P 2, p. 4: AT 8A, p. 42; CSM 1, p. 224.

The thought-experiment that Descartes presents here is designed to show that hardness is relative to our bodily sense of touch. If we conceive of bodies in the absence of touch, using our pure intellect, we will make no mention of hardness. Hardness means only 'that the external bodies possess what is required to bring it about that our nerves excite in the soul the sensations of hardness . . . '.[10] Our perception of hardness is, at best, only indirectly informative of the nature of body as understood by pure intellection. The thought-experiment is therefore a way of appealing to the reader to exclude such sensual perception from our conception of material substance. As Descartes later affirms, 'a body can retain its whole bodily nature without being soft or hard [*durus*] or cold or hot to the senses—indeed without having any perceptible quality'.[11]

But while these reflections may help us appreciate the sensual component in the quality of hardness, we are still owed an explanation of how our perception of hardness emerges from the underlying nature of corporeal substance. To this end, Descartes appeals to the action of particles—an explanation which avoids circularity by making no appeal to *hardness* in those particles. He argues that the cause of tangible hardness is an absence of internal motion in the particles that make up a given body.[12] It is this mutual rest within that not only makes it difficult for our hands to divide hard objects, but also makes it painful to collide with them, and generally means they maintain their shape through our various interactions with them.

Descartes's explanation of hardness may look vacuous, and it even came to be seen as such by some of his followers—a point I shall return to below. By appealing to internal rest, he may be thought to be merely re-describing hardness, not explaining it. In the context of the principles of his physics, however, his account of hardness is an informative one. Descartes is treating the phenomenon of hardness as a special case of the law of inertia according to which an object has a natural tendency to remain in the same state of rest or motion.[13] Descartes's principle of inertia, it will be remembered, only concerns the *quantity* of motion (speed times size), not its direction, or 'determination'.[14] Thus, where there is already internal motion in bodies (fluids), division is easy because we do not need to overcome the inertia of the particles that make it up, but only to change their direction of motion. In hard bodies, on the other hand, where division requires the introduction of new motion, an external force sufficient to overcome the inertia of the resting particles will be needed.[15]

Whatever the merits of Descartes's explanation of hardness—and of the other degrees of bodily cohesion—it led him to put hardness and softness in the same category as other sensible qualities such as colour, heat, and cold. These qualities were

[10] P 4, p. 191: AT 8A, p. 318; CSM 1, p. 282.

[11] To More, 5. 2. 1649: AT 5, p. 268; CSMK, p. 360. See also P 2, p. 11: AT 8A, pp. 46–7; CSM 1, pp. 227–8.

[12] P 2, pp. 54–63: AT 8A, pp. 70–9; CSM 1, pp. 245–7.

[13] P 2, p. 37: AT 8A, pp. 62–3; CSM 1, pp. 240–1.

[14] For Descartes's distinction between the opposition of rest and motion on the one hand and different directions of motion on the other see P 2, p. 44: AT 8A, p. 67; CSM 1, p. 244.

[15] I have further explored the details of Descartes's view of hardness and bodily cohesion in my 'Locke's Account of Cohesion'.

not to be attributed to the intrinsic nature of body. Instead, they were held to inform us of whether objects 'are beneficial or harmful to man's composite nature'.[16] Those who took these qualities to really exist in external objects were charged with committing what Cartesians saw as the cardinal philosophical sin—that of placing unthinking trust in the senses.

4. The Distinction between Impenetrability and Hardness in Descartes

At this point it becomes important to distinguish between impenetrability and hardness in Descartes's thought. While hardness is dealt with explicitly in his *Principles*, Descartes's view of impenetrability is much less easy to discern and can be easily overlooked. In fact, it only comes out clearly when polemical pressure is brought to bear on his philosophy of body by the Cambridge Platonist Henry More. In the correspondence with More, we find Descartes willing to use the term 'impenetrability' in the traditional scholastic sense; that is, as referring to the property of matter that excludes the possibility of two bodies being in the same place at the same time. Impenetrability, in this sense, is not to be contrasted with divisibility as there is no obvious reason why an impenetrable body should not be separated or divided. The contrast is, instead, with *penetratio* by which bodies would somehow overlap or merge, so as to share the same spatial coordinates.

Descartes affirms to More that all extended things—and therefore all bodies—are impenetrable. This is because extension means having *partes extra partes*. These parts are, by their very nature, external to one another, and cannot therefore coexist in the same place at the same time. In his second letter to More, Descartes defends the necessary impenetrability of extended things:

> it is impossible to conceive of one part of an extended thing penetrating another equal part without thereby understanding that half the total extension is taken away or annihilated; but what is annihilated does not penetrate anything else; and, so, in my opinion, it is established that impenetrability [*impenetrabilitas*] belongs to the essence of extension and not to that of anything else.[17]

Impenetrability for Descartes is a property quite distinct from hardness. Unlike hardness, impenetrability is a necessary and universal characteristic of extension in three dimensions. It is not an emergent quality that might be explained by the relative motion of constituent particles, or by their effect on our physical bodies. Nor is the impenetrability of body something that we learn by sense experience. Rather, as the passage just quoted attests, Descartes took the universal impenetrability of body to be a conceptual

[16] P 2, pp. 3–4: AT 8A, pp. 41–3; CSM 1, pp. 224–5.

[17] AT 5, p. 342; CSMK, p. 372 Locke owned a collection of Descartes's letters in his library that included the letters to More (see LL 101).

necessity. To deny impenetrability would be an affront to reason: it would amount to the nonsensical claim that a part of space could be temporarily, or even permanently, annihilated. As Descartes explains in the same letter to More, if some part of space were thought to penetrate another, then a part of space would thereby be destroyed and cease to be. This would be patently absurd and so he concludes that 'impenetrability must be admitted in every space' (*impenetrabilitas in omni spatio est admittenda*).[18]

Whether or not Descartes's treatment of impenetrability in these letters to More, which were written less than two years before the Frenchman's death, complicates his famous claim that the essence of bodily substance is extension in length, breadth, and depth, or whether (as seems more likely) he had always tacitly held that extension brings with it the impenetrability of its parts, is a question that lies beyond the scope of this chapter.[19] The important point for our purposes is that a fundamental division between hardness and impenetrability came to characterize Descartes's legacy. Impenetrability was a quality common to all extended things, but hardness was a quality displayed only by some bodies, and then in relation to us. While impenetrability was deducible from the nature of body itself, hardness called for a hypothesis to explain it. And, as we have seen, Descartes took pains to provide such a hypothesis by appealing to the mutual rest of the constituent particles.

5. Locke on Hardness and Impenetrability

We do not have to look far in Locke's published *Essay* to find a parallel distinction between hardness and impenetrability. Locke, in the chapter 'On Solidity', follows Descartes in treating hardness as a quality that is secondary, and relative to bodily sensation.

> Hard and Soft are Names that we give to things, only in relation to the Constitutions of our own Bodies; that being generally call'd hard by us, which will put us to Pain, sooner than change Figure by the pressure of any part of our Bodies; and that, on the contrary, soft, which changes the Situation of its parts upon an easie, and unpainful touch.[20]

Impenetrability, on the other hand, is universal to bodies and 'inseparably inherent in Body, where-ever, or however modified'. Locke, it is true, prefers the term 'solidity' to

[18] AT 5, p. 342; CSMK 373.

[19] Descartes had mentioned impenetrability in passing in the Sixth Replies (AT 7, p. 442; CSM 2, p. 298). For an informative discussions of the issue of impenetrability in Descartes's thought see Garber, *Descartes' Metaphysical Physics*, pp. 144–8, and Reid, *The Metaphysics of Henry More*, pp. 63–70. Reid takes impenetrability to be a 'secondary attribute' of matter, for Descartes, deducible from the primary attribute of extension, and he argues that Descartes thought mention of this secondary attribute, prior to the exchange with More, was 'simply unnecessary' (p. 64). Garber, however, notes that impenetrability is not applicable to 'the totality of corporeal substance', but only when its parts are considered (p. 148). If Garber is right, then Reid's assertion that impenetrability is an *attribute* of material substance is problematic, although impenetrability would remain a necessary characteristic of any of the bodies that make up the totality of matter.

[20] E II. iv. 4, p. 125.

'impenetrability', mainly because the former term 'carries something more of the positive in it'. But he accepts that the terms may be used interchangeably when he adds—with a tolerant shrug—that 'if any one think it better to call it *Impenetrability*, he has my consent'. Solidity or impenetrability is the property of bodies by means of which they fill space and it is the 'resistance which we find in Body, to the entrance of any other Body into the Place it possesses, till it has left it'. While solidity means the 'utter Exclusion of other Bodies out of the space' the body possesses, hardness is the 'firm Cohesion of the parts of Matter, making up masses of sensible bulk'.[21]

Locke, unlike Descartes, presents a series of empirical considerations to back up the universal ascription of impenetrability to bodies, drawing both on experimentation by the 'virtuosi' and on everyday observation. We are told of the Florentine experiment in which a golden globe filled with water began to perspire through 'pores' in the metal as the globe was forcibly compressed, confirming that the water inside the globe was solid or impenetrable.[22] We are also reminded of how we continually experience the resistance of body as the ground supports us[23]—or, as Locke puts it in Draft C, 'we neither goe sit stand nor lie but it Constantly affects us'.[24]

Many of the examples that Locke provides of solidity are actually—and quite intentionally—ones involving *soft* bodies. Locke was keenly aware of the potential confusion of solidity (and thus impenetrability) with hardness. So, for example, he writes, 'He that shall fill a *yielding soft* Body well with Air or Water, will quickly find its Resistance: And *he that thinks, that nothing but Bodies, that are hard*, can keep his Hands from approaching one another, may be pleased to make a trial, with the Air inclosed in a Football.'[25]

But how, one might ask, can Locke prove that *all* bodies are impenetrable in the framework of his empiricist system? Is this not an inductive generalization that goes well beyond any a posteriori findings about the bodies we come into contact with? Locke was no doubt aware of this difficulty and his response to it is to follow Descartes in treating the impenetrability of body as ultimately a conceptual matter. In Locke's understanding, once we treat a body as *filling* a place equal to the space within its surfaces it becomes a self-evident proposition '*That two Bodies cannot be in the same place*'.[26] The exclusion of penetrability is known by an 'immediate perception' of the mind, Locke writes, meaning a simple act of intuition about what it means for matter to occupy space.

The nature of this intuition becomes somewhat clearer when Locke expands on the conceptual consideration in his famous chapter 'Of Identity and Diversity' added to the second edition of the *Essay* of 1694. Here Locke argues for the universality of impenetrability by a *reductio ad absurdum*. If we allow that two particles of matter can occupy the same space, then we must also allow that 'all Bodies may be in one place'.[27]

[21] E II. iv. 1–4, pp. 123–5. [22] E II. iv. 4, pp. 125–6. [23] E II. iv. 1, 123.
[24] Draft C, II. iii. 2. [25] E II. iv. 4, p. 125, my emphases. [26] E IV. vii. 5, p. 594.
[27] E II. xxvii. 2, p. 329.

The possibility of penetration thus has the 'ridiculous' consequence that the whole material world might merge at a single point—a consequence that, once granted, would destroy the very distinction between the identity and diversity of bodies. This *reductio* depends, of course, on assuming the homogeneity of matter, because it tacitly assumes that what we allow to one particle we must allow to all particles. However, the homogeneity thesis was a commonplace in early modern theories of matter and therefore the argument was not one that could be easily dismissed by his contemporaries.

So, while Locke asserts that our idea of solidity is to be originally gained from our sense experience of touch, he treats the universal solidity or impenetrability of matter as known by reflection on the intuitively absurd consequence of physical penetration which would allow a body to be annihilated by other bodies. It is this thought that, I think, lies behind Locke's claim that:

> All the Bodies in the World, pressing a drop of Water on all sides, will never be able to overcome the Resistance, which it will make, as soft as it is, to their approaching one another, till it be removed out of their way...[28]

If this drop of water were to yield under the pressure, and to pass out of existence, so might any particle of matter moving into its place disappear, permitting the progressive disappearance of the material world itself. So Locke, like Descartes, held that the universal ascription of impenetrability was required if we were to prevent the absurd scenario of the material world, or any of its parts, being annihilated by natural means.

6. The Scope of Locke's Impenetrability

Locke's distinction between hardness and impenetrability parallels that of Descartes. The parallel is to be found not only in the relativity to sensation characteristic of hardness, but also in the conceptual or demonstrative status of the assertion that all bodies are impenetrable. But, despite these similarities, there is an obvious difference in scope. Descartes's geometrical understanding of matter, *res extensa*, means that 'there is no real difference between space and corporeal substance',[29] so that when he says all parts of matter are impenetrable he is talking of extension itself, not just material things in distinction to space. Locke, on the other hand, seeks to maintain the intuitive distinction between space and the matter it may contain. He therefore treats impenetrability, the defining property of material things, to be predicable *only* of material things—that is, of material things in distinction to the space they occupy.

This is, of course, a significant difference, and Locke makes much of it. His aim, after all, is to defend a view of matter and body that is gained from sense experience, and to turn his back on what he saw as the unacceptable consequences of the Cartesian geometrical conception, based on pure intellection. His attack on Cartesian *res extensa*

[28] E II. iv. 3, p. 124.

[29] P 2, p. 11: AT 8A, p. 46; CSM 1, pp. 227–8.

is already a notable part of Draft A of the *Essay* of 1671, where he wrote that to equate body with extension was to play an empty verbal game that led to various absurdities including a spurious a priori demonstration of the impossibility of the vacuum.[30]

But how does Locke limit impenetrability to material bodies and resist the Cartesian argument for attributing impenetrability to space itself? He does this by questioning the conception of the parts of space that he takes to lie behind the Cartesian view. Locke argues that space itself has no *separable* or *movable* parts because it has no surfaces ('superficies'). Surfaces are to be predicated only of solid bodies. This means that even the notional possibility of a part of space being removed and placed elsewhere is actually quite unthinkable. Only a separable solid part can be conceived to occupy a place and thus to potentially invade the place occupied by another part.[31] Thus while we must attribute impenetrability to all the separable parts of solid bodies, because separability and impenetrability mutually entail one another, Locke thought it quite unjustified to attribute impenetrability to space itself.

So the difference of scope is really a result of Locke's differing understanding of the nature of extension and space. He agrees with Descartes that only insofar as it has divisible parts can extension be characterized as impenetrable. Indeed, there are signs that for Locke such divisible, bodily, extension is the proper meaning of the very term 'extension'. At one point he undertakes to refer to indivisible space, when understood in abstraction from body, as 'expansion'.[32] Although this terminological undertaking is soon forgotten, it would have enabled Locke to capture the conceptual connection between 'extension' and impenetrability just as readily as Descartes, since extension for both of them would be 'that which is considered as divisible into parts'. We might also note that Descartes himself, as Daniel Garber has stressed, characterizes only the *parts* of extension—not its totality—as impenetrable.[33] Thus his position, like Locke's after him, asserts the conceptual connection between extension and impenetrability only insofar as divisible parts are considered.

7. Locke and Atomism

We have now arrived at a vantage point from which we can assess how Locke's approach to body, with its distinction between impenetrability and hardness, is related to the Epicurean view. There is no doubt that Locke shared many crucial doctrinal viewpoints with the atomists. Most importantly, he accepted both the conceivability and the existence of the vacuum. Indeed, one of his main arguments for vacuuism was the traditional atomist one that without empty space there could be no movement—Locke, like Lucretius before him, argued that a plenum would mean eternal gridlock.[34]

[30] *Drafts*, p. 45. [31] E II. viii. 13–15, pp. 172–3. [32] E II. xv. 1, p. 196.

[33] Garber, *Descartes' Metaphysical Physics*, p. 148.

[34] E II. xiii. 22, pp. 177–8. The same argument is to be found in Lucretius at 1. 335–45 where it is argued that in a plenum 'nothing could advance, since nothing could make a start of yielding'. The same thought

From the lexical point of view, of course, there is also a similarity between Locke's English term 'solidity' and the Latin term *soliditas* which Lucretius used to characterize the indivisible atoms. It is not surprising, then, that some commentators have been tempted to follow Leibniz and to regard Locke's theory of matter as essentially Epicurean, and to treat solidity as the insuperable hardness of the Epicurean atoms.

One commentator who takes this line is Peter Alexander. Alexander thinks of solid corpuscles as indivisible because they have no interstices or 'pores', and thus treats Locke's corpuscularianism as a form of atomism.[35] But this temptation should be resisted. Locke's use of the term 'solidity' is quite different from the atomist use. For Locke it has no necessary connection with indivisibility. The many examples he gives of solid objects—including liquids and gases—are all, in practice or in principle, divisible into parts. Locke never assumes there must be an endpoint to physical division. His different discussions of divisibility invariably leave room for the process to go on indefinitely.[36]

Indeed, later in the *Essay*, Locke will imply that it is the very solidity of physical objects that makes possible their division. He refers to the division of objects by 'intrusion', where one object (e.g. a wedge) is used to drive apart another object (e.g. a log), arguing that this kind of division is intelligible to us only because we can see how 'the separation of the Parts of one Body' is necessitated by 'the intrusion of another' solid body.[37] His thought here seems to be that because all bodies are impenetrable, the intruder cannot be absorbed into the body it enters, but must force itself between its parts, dividing them from one another. In other words, solidity, far from precluding division, actually makes it possible.

At this point, the atomist interpreter might claim that Locke—whether he says it or not—is presupposing indivisible particles in talking of the impenetrability of soft things and of fluids. He is assuming that they are constituted by perfectly hard atoms that resist all division. Perhaps he would endorse the atomist claim that only thus can we explain how the hardness and softness of perceivable bodies emerge. This claim was grounded by the atomist on the fact that while a world of soft particles would give us no way of explaining how hard bodies could ever arise, hard particles may be imagined to form larger soft bodies by being interspersed with vacua.[38]

can be traced back to the Eleatics when Melissus of Samos argues that because there is no void in things they cannot give way to allow motion (Melissus B7.7).

[35] Alexander, 'Solidity and Elasticity', pp. 144–6 and p. 155.

[36] See, for example, E II. viii. 9, pp. 134–5. The distinction between Locke's concept of solidity and the atomist one of indivisibility is nicely made and illustrated by Wilson in 'Locke's Primary Qualities', pp. 105–6.

[37] E IV. iii. 13, p. 545.

[38] This argument was, of course, a traditional atomist one that was stated, for example, by François Bernier, Locke's friend, as follows: 'posant la solidité des Atomes, on peut non seulement rendre raison de la solidité qui est dans les choses, mais aussi de la Mollesse; en ce que les choses sont dures acause qu'elles sont formées d'Atomes qui ont de la dureté, et molles en ce que bien qu'elles soient formées d'Atomes durs, elles ont neamoins des Vuides interceptez…' (Bernier, *Abrégé*, volume 3, book 1, p. 112).

Could Locke have been covertly endorsing this atomist line of thought? I think not. By the time of the first edition of the *Essay* it would be at variance with his carefully made distinction between the solidity and the cohesion of bodies. Hardness was, lest we forget, 'firm Cohesion',[39] and cohesion was something quite different from solidity or impenetrability. The cohesion of matter, of which hardness represented an especially high degree, was quite inexplicable to us, despite our attributing solidity to matter. Indeed, Locke was ready to pursue the question of cohesion down to the smallest particle of matter conceivable and demand an account of it there:

> He that could find the Bonds, that tie these heaps of loose little Bodies together so firmly; he that could make known the Cement, that makes them stick so fast one to another, would discover a great, and yet unknown Secret: And yet when that was done, would he be far enough from making the extension of Body...intelligible, till he could shew wherein consisted the union, or consolidation of the parts of those Bonds, or of that Cement, or of the least Particle of Matter that exists.[40]

Locke here rejects the atomist assumption that invincibly hard particles provide a solution to the problem of cohesion. Postulating them merely transfers the problem of cohesion to a level of imperceptible minuteness.

8. Hardness and the Distinctive Agnosticism of Locke's Philosophy of Body

I began by saying that Locke's agnosticism about the essence of matter is his distinctive philosophical conclusion in this area. It now remains for us to indicate how his acceptance of the Cartesian treatment of hardness played a role in this agnostic stance which led him to oppose not only Descartes's *res extensa* doctrine, but also the atomist dogma that indivisibility is essential to body. To this end, we must say something more about the problem of the cohesion of matter which we mentioned in Section 7.

The atomists did not recognize any serious or foundational problem with cohesion. The atoms possessed internal cohesion by virtue of their inner plenitude—or 'defect of inanity', as Walter Charleton put it—which was thought to grant them indivisible hardness.[41] This then made it possible to explain the relative cohesion of larger bodies by the atoms becoming more or less entangled with one another. The simplest version of this would be a hook-and-eye theory, allowing the parts of the hard atoms to become interlocked so as to cause a firmness in larger bodies. Many sophisticated, multi-layered, versions of this model had been developed by the time Locke was writing.[42] But Locke found no refuge in them. His solidity, as we have shown, was carefully

[39] E II. iv. 4, p. 125. [40] E II. xxiii. 26, p. 310.

[41] Charleton, *Physiologia*, p. 330. Charleton goes on to say that 'the intermixture of inanity' is 'the Cause of all Divisibility'.

[42] A thorough treatment of these theories is to be found in Millington, 'Theories of Cohesion'.

distinguished from hardness and amounted only to the resistance involved in filling space. Solidity in this sense was no less characteristic of soft bodies and fluids than of hard ones. A perfectly solid body would not be rendered hard and coherent simply by lacking internal vacua. A foundational problem of cohesion therefore presented itself to Locke, which the atomists had never confronted—the problem of how *any* body, including one of the putative atoms, was able to cohere.

This problem may not yet to have struck Locke in the early Drafts of the *Essay*. In Draft B, for example, he appears satisfied with an explanation of hardness along mechanistic lines, appealing to entanglement of parts which are themselves (presumably) already hard.[43] But, by the time of Draft C of 1685, the foundational problem of the cohesion of body now clearly troubled Locke and he dwelt on it in his chapter on solidity.[44] This passage was later merged into the chapter on substance in the first edition of the *Essay* of 1690,[45] where the problem of cohesion helped justify his agnosticism about the real essence of material substance. If we cannot begin to understand how the cohesion of body comes about, he now argued, how can we claim to understand the underlying nature of matter?

Why did Locke become convinced that the problem of cohesion was irresolvable? One reason is that he came to see Cartesian explanations of cohesion as hopelessly circular. Descartes's mutual rest explanation had, perhaps, already died a death by the 1680s when even his followers were looking to other explanations of cohesion. Malebranche, for example, had roundly dismissed Descartes's account of cohesion as founded on the 'false premise that rest has force'.[46] There was now a growing tendency in Cartesian circles to turn to the ether (*la matière subtile*), which had already been put to use by Descartes in explaining a host of natural phenomena, including magnetism and gravitation. The hope was that cohesion might be explained by the pressure of this subtle fluid pushing into the pores and cavities of bodies from without. Indeed, Descartes himself had already appealed to the ether in a similar way when accounting for the narrower quality of rigidity.[47] Such an approach was developed in detail not just by Malebranche, but also by Jakob Bernoulli in his *Dissertatio de Gravitate Aetheris* (1683), a copy of which Locke purchased soon after its publication when in exile in Holland.[48] Locke's main contention against this aetherial explanation is that the particles of the ether must themselves cohere if they are to exert pressure on larger bodies, demanding another—yet more subtle—ether, and so on ad infinitum. This exposure of a vicious circularity in the new Cartesian explanation of the cohesion of body came to represent, for Locke, the inevitable failing of any attempt to account for the firmness

[43] See *Drafts*, p. 156 where he talks of hardness in water being produced in 'the way a joyner puts severall peices of wood togeather to make a box or table, which by tenants nailes & pins we well enough perceive how it is made to hang togeather'.

[44] Draft C II. xxi. 7–10. [45] E II. xxiii. 23–7, pp. 308–11.

[46] Malebranche, *The Search after Truth* 6. 2. 9, pp. 510–26. [47] P 4, p. 132; AT 8A, p. 274.

[48] Bernoulli was mentioned by name in this connection in Locke's *Abrégé*, see pp. 76–7.

and hardness of bodies within the mechanical system, and it confirmed his view that a knowledge of the real essence of material things was quite beyond our powers.

9. Conclusion

Locke's agnosticism about the essence of matter was already a part of his thinking in the first drafts of the *Essay* that he composed in 1671.[49] Initially, he based his agnostic stance on his empiricism, which limits our knowledge of bodies to the sensible qualities inherent in material substance, and excludes any perception of their substratum.[50] But his adoption of the Cartesian distinction between impenetrability and hardness gradually led him to develop a more elaborate argument for this agnosticism, which is presented in Draft C and in the *Essay* itself. If the impenetrability or solidity of bodies could not secure their cohesion, we cannot begin to explain how bodies hold together. Thus there is a mystery in our very conception of material things.

Locke's close attention to the travails of the Cartesians in attempting to explain the hardness or cohesion of body confirmed him in his scepticism. He could neither accept the dogmatic assumption of the Epicureans that there were absolutely hard atoms that simply had cohesion and indivisibility granted to them, nor could he accept the Cartesian hypotheses of internal rest or external aetherial pressure that sought to account for the emergence of hardness. As a result, the problem of cohesion becomes the primary reason for Locke's distinctive sceptical stance towards the essence of matter in the published *Essay*. Locke's agnosticism—at odds with the thought of both Descartes and the atomists—had been allowed to develop and gain justification because of the Cartesian element in Locke's philosophy of body.[51]

References

Alexander, Peter. 'Solidity and Elasticity in the Seventeenth Century', in G. A. J. Rogers, ed., *Locke's Philosophy: Content and Context*. Oxford: Oxford University Press, 1994: 143–64.

Bernier, François. *Abrégé de La Philosophie de Gassendi* (1684). 8 vols. Paris: Fayard, 1992.

Bernoulli, Jakob. *Dissertatio de Gravitate Aetheris*. Amsterdam, 1683.

Charleton, Walter. *Physiologia Epicuro-Gassendo-Charltoniana* (1654). Introd. Robert Kargon. London: Johnson Reprint, 1966.

Garber, Daniel. *Descartes' Metaphysical Physics*. Chicago: University of Chicago Press, 1992.

Hill, James. 'Locke's Account of Cohesion and its Philosophical Significance', in *John Locke: Critical Assessments*, II, ed. Peter Anstey. 4 vols. Oxford: Routledge, 2006: vol. 3, 145–65.

[49] *Drafts*, p. 2.

[50] See also *Drafts*, p. 130 (Draft B §19) where the problem is stated as our being ignorant of 'the substratum to those simple Ideas we have from without'.

[51] Funding for work on this article has been kindly provided by the Grant Agency of the Czech Republic (project 16-12624S).

Leibniz, Gottfried. *New Essays on the Human Understanding* (*Nouveaux essais sur l'entendement humain*), trans. P. Remnant and J. Bennett. Cambridge: Cambridge University Press, 1986.

Lennon, Thomas. *The Battle of the Gods and Giants: The Legacies of Descartes and Gassendi, 1655–1715*. Princeton: Princeton University Press, 1993.

Locke, John. *Abrégé* or 'Extrait d'un Livre Anglois qui n'est pas encore publié', in *Bibliotheque Universelle et Historique*, ed. J. Le Clerc, Amsterdam, 1688: vol. 8, 49–142.

Locke, John. *Drafts for the* Essay Concerning Human Understanding, ed. Peter Nidditch and G. A. J. Rogers. Oxford: Clarendon, 1990.

Lucretius. *De Rerum Natura*, trans. C. Bailey. Oxford: Oxford University Press, 1947.

Malebranche, Nicolas. *The Search after Truth* (*De la recherche de la vérité, 1674–5*), trans. T. M. Lennon and P. J. Olscamp. Cambridge: Cambridge University Press, 1997.

Millington, E. C. 'Theories of Cohesion in the Seventeenth Century', *Annals of Science*, 5 (1941–7), 253–69.

Milton, J. R. 'Locke at Oxford', in G. A. J. Rogers, ed., *Locke's Philosophy: Content and Context*. Oxford: Oxford University Press, 1994: 29–47.

Milton, J. R. 'Locke, Medicine and the Mechanical Philosophy', *British Journal for the History of Philosophy*, 9/2 (2001), 221–43.

Reid, Jasper. *The Metaphysics of Henry More*. New York: Springer, 2012.

Walmsley, Jonathan, 'The Development of Locke's Mechanism in the Drafts of the Essay', *British Journal of the History of Philosophy*, 11/3 (2003), 417–49.

Wilson, Robert A. 'Locke's Primary Qualities', in *John Locke: Critical Assessments*, II, ed. Peter Anstey. 4 vols. Oxford: Routledge, 2006: vol. 3, 89–121.

Woolhouse, Roger. 'Lady Masham's Account of Locke', *Locke Studies* 3 (2003), 167–93.

4

Are Body and Extension the Same Thing?

Locke versus Descartes (versus More)

Lisa Downing

In the course of considering the simple modes of space in Book II of the *Essay*, Locke notes that 'There are some that would persuade us, that *Body and Extension are the same thing*' (E II. xiii. 11). Those would be the Cartesians, of course, though Locke, as is typical of him, does not name the sect. Descartes held that extension is the principle attribute or essence of body (corporeal substance) and that in some sense body is *nothing more than* mere extension, for body is just extension and its modifications.[1] Notoriously, this provides Descartes with a very short argument for the impossibility of a vacuum: Suppose there were a bit of empty space. That space would have to be extended, but extension suffices for corporeality, ergo it would actually be body. Space, then, is necessarily full, a plenum.

Locke is suspicious of both the conclusion and the shortness of the argument.[2] I focus here, however, on his attempts to prove in E II. xiii. 11–14 that we have distinct ideas of body and of extension. My goal is both to evaluate this anti-Cartesian foray, and to use it to reflect on some intriguing and abstruse elements of Descartes's ontology of body. I will also use it to motivate some general reflections on Locke's relation to Descartes.

1. Locke contra Descartes on Body versus Space: E II. xiii. 11–14

First, though, we might want to reflect on the difference between the title question (are body and extension different things), and the question of the distinctness of our

[1] Better: all modes of body are ways of being an extended substance.

[2] I consider Locke's arguments in E II. xiii. 21–3 against the Cartesian plenum in Downing, 'Locke and Descartes,' pp. 111–15. These arguments too are finely tuned against Descartes's actual positions and responsive to Descartes's correspondence with More.

ideas. Locke consistently maintains that the core of his dispute with Descartes about body and space concerns merely our ideas.

The focus on ideas has more than one explanation. The most obvious, perhaps, is that Locke is in the middle of Book II of his *Essay concerning Human Understanding*, and thus is engaged in an inventory and analysis of our ideas, so what he needs to defend is a thesis about the content of our ideas. The other, of course, is that by identifying the ideas of extension and body, Descartes has foreclosed on the *possibility* of empty space, which Locke wants to reopen. If he can show that the idea of body is not the same as or included in the idea of extension, then there is no conceptual impossibility in extension existing without body existing.[3]

In E II. iv. 5, Locke remarks pointedly that if the Cartesians are serious, no one can communicate with them, for they are abusing language. This charge is clarified in E II. xiii. 11, as Locke turns from his account of our spatial ideas to his critique of its Cartesian competitor. Here Locke complains that they either 'change the Signification of Words, which I would not suspect them of, they having so severely condemned the Philosophy of others, because it hath been too much placed in the uncertain meaning, or deceitful obscurity of doubtful or insignificant Terms' or they 'confound very different ideas together'; that is, either they are simply abusing language by covertly using words in non-standard ways, or they are conflating two distinct ideas. Locke is clear here that the two distinct ideas, which should not be confounded, are that of extension and that of body, and that it is solidity that makes the difference between them. '*Body*', he tells us, means 'something that is solid, and extended, whose parts are separable and movable different ways', while 'Extension' means 'only the Space that lies between the Extremities of those solid coherent Parts, and which is possessed by them' (E II. xiii. 11).

The distinction between the ideas associated with the terms 'body' and 'extension' should, Locke charges, be plain 'to every Man's own Thoughts'. Before arguing for this purportedly obvious conclusion, he pauses to address briefly what he takes to be a Cartesian objection to his 'distinct ideas' claim. He admits some conceptual connection between solidity and extension, that solidity can't be conceived without extension, but draws two comparisons:

> Motion can neither be, nor be conceived without Space; and yet Motion is not Space, nor Space Motion: Space can exist without it, and they are very distinct *Ideas*; and so, I think, are those of Space and Solidity. (E II. xiii. 11)

> And if it be a Reason to prove, that Spirit is different from Body, because it includes not the *Idea* of Extension in it; the same Reason will be as valid, I suppose, to prove, that *Space is not Body*, because it includes not the *Idea* of Solidity in it; *Space and Solidity* being *as distinct Ideas*, as Thinking and Extension, and as wholly separable in the Mind one from another. (E II. xiii. 11)

The second comparison may strike us as obtuse, since of course Descartes claims that a two-way conceptual separability holds between extension and thought, so they are

[3] Locke typically treats 'space' and 'extension' as interchangeable, and he endorses this in E II. v and E II. xiii. 3.

'wholly separable' where extension and solidity are not.[4] One might suspect that Locke's own views lurk in the background here, since in fact he thinks that while extension does not require thought, we do conceive of thinking things as spatially located, and beyond that their relation to extension is unclear. Thus, in Locke's view, in both cases only one-way separability can readily be defended, but that suffices to establish the distinctness of the ideas.

Locke proceeds to give three numbered reasons why ''tis evident' that *Body* and *Extension* are two distinct ideas. For the first reason, Locke simply asserts that '*Extension* includes no Solidity, nor resistance to the Motion of *Body*, as Body does'. The second and third reasons focus on the separability and mobility of the parts of body, which, Locke holds, contrasts with the inseparability and immobility of the parts of space:

> *Secondly*, The Parts of pure Space are inseparable one from the other; so that the Continuity cannot be separated neither really, nor mentally. For I demand of any one, to remove any part of it from another, with which it is continued, even so much as in Thought. (E II. xiii. 13)
>
> *Thirdly*, The parts of pure *Space* are immovable, which follows from their inseparability: *Motion* being nothing but change of distance between any two things: But this cannot be between Parts that are inseparable... (E II. xiii. 14)

It is not at all clear that there are two distinct arguments here, as Locke seems to assert both that mobility presupposes separability and that separability presupposes mobility:

> To divide and separate actually, is, as I think, by removing the parts one from another, to make two superficies, where before there was a continuity; and to divide mentally, is to make in the mind two superficies, where before there was a continuity... (E II. xiii. 13)

Nevertheless, Locke's intended contrast is clear enough.

If we focus on the first argument, we may be tempted to conclude that Locke and Descartes really are just arguing about words and that the word they disagree about is 'body'. They disagree, it seems, in that Descartes stipulates that having extension suffices to have body, while Locke maintains that an extra ingredient is required, namely, solidity. This, however, would be to crucially misread the dialectic. For it turns out that Descartes holds that all bodies are extended and impenetrable, just as Locke holds that all bodies are extended and solid.[5] Furthermore, Locke equates solidity and impenetrability:

> That which thus hinders the approach of two Bodies, when they are moving towards another, I call *Solidity*.... if anyone will think it better to call it *Impenetrability*, he has my Consent.

[4] That is, extension can be conceived without thought, and thought without extension; thus (according to Descartes) they exhibit two-way separability. But, in contrast, while extension can be conceived without solidity, solidity cannot be conceived without extension, so only one-way separability obtains between them.

[5] In discussing Descartes, Des Chene (*Physiologia*, p. 379) distinguishes impenetrability from hardness and solidity. Hardness, as he reads it, is a body's power, derived from cohesion, to resist our bodies, while solidity is a body's power, derived from cohesion, to resist division into parts. I'm not sure what's to be gained in seeing these as two distinct powers. In any case, the important point from our perspective is that impenetrability, unlike hardness or solidity in these senses, is an intrinsic quality of all matter. And this impenetrability is what Locke calls 'solidity'.

Only I have thought the Term *Solidity* the more proper to express this *Idea*, not only because of its vulgar use in that Sense, but also, because it carries some thing more of positive in it than *Impenetrability*, which is negative, and is, perhaps, more a consequence of *Solidity*, than *Solidity* itself. (E II. iv. 1)[6]

That Descartes and Locke held overlapping views here is further suggested by the fact that they both distinguish this universal quality of body, that of solidity/impenetrability, from *hardness*, which both declare to be something different, and something dependent on our sense of touch:

For as regards hardness, our sensation tells us no more than that the parts of a hard body resist the motion of our hands when they come into contact with them.

(P 2, p. 4; AT 8A, p. 42; CSM 1, p. 224)

And indeed, Hard and Soft are Names that we give to things, only in relation to the Constitutions of our own Bodies; that being generally call'd hard by us, which will put us to Pain, sooner than change Figure by the pressure of any part of our Bodies; and that on the contrary, soft, which changes the Situation of its parts upon an easie, and unpainful touch. (E II. iv. 4)

Locke and Descartes, then, are in near-perfect agreement on *body*, at least at this level of analysis. They both hold that all bodies are extended and solid/impenetrable. Where they disagree, then, must be with respect to *space*. Locke holds that space is merely extended but not solid. Descartes, however, holds that anything extended is *ipso facto* impenetrable, for extension *entails* impenetrability. Impenetrability is thus, in the technical Aristotelian sense, a *property* of all extended things:

Now tangibility or impenetrability in body is something like the ability to laugh in man; according to the common rules of logic it is a 'property of the fourth kind', and not a true and essential differentia such as I claim extension to be. (AT 5, p. 269; CSMK, p. 361)[7]

2. Descartes on Impenetrability and Extension

Why does Descartes hold that extension follows from impenetrability? It is clear enough why he needs to hold it. He needs to hold it for his metaphysical physics. As Locke eloquently pointed out, upon impenetrability/solidity 'depends its filling of Space, its Contact, Impulse, and Communication of Motion upon Impulse' (E II. xiii. 11). Descartes needs his bodies to be impenetrable, and because of his metaphysical

[6] What Locke is getting at here is that impenetrability expresses the fact that two bodies can't overlap, can't occupy the same space, whereas solidity is supposed to denote that quality which prevents them from occupying the same space. That is, despite the fact that Locke's characterizations of solidity often sound dispositional, solidity itself is supposed to be a categorical property.

[7] *Risibility* is the classic Aristotelian example of a property in this sense, a quality that follows from the essence (in this case, *rational animal*), but isn't the essence. The fourfold division can be found in Porphyry. A property of the fourth kind belongs to all members of the species at all times and only to members of the species. See Des Chene, *Physiologia*, p. 380.

commitments, he needs that not to be a further quality, over and above extension, which is *the* attribute of body, and thus suffices for something to be body.

In his published works, Descartes seems to assume without comment that bodies are impenetrable, but he supports this with argument in his correspondence with Henry More. It is one theme of this chapter that we should see this correspondence as lying behind Locke's treatment of Descartes on body and space. Descartes gives there what I think we should regard as two versions of one argument.

> For in a space—even an imaginary and empty space—everyone readily imagines various parts of determinate size and shape; and some of the parts can be transferred in imagination to the place of others, but no two of them can in any way be conceived as compenetrating each other at the same time in one and the same place, since it is contradictory for this to happen without some part of space being removed. (AT 5, p. 271; CSMK, p. 362)

> ... it is impossible to conceive of one part of an extended thing penetrating another equal part without thereby understanding that half the total extension is taken away or annihilated; but what is annihilated does not penetrate anything else; and so, in my opinion, it is established that impenetrability belongs to the essence of extension and not to that of anything else. (AT 5, p. 342; CSMK, p. 372)

Descartes's argumentation here is potentially misleading. It may sound as if he holds that two extended things might move together, and apparently interpenetrate, but that what must happen in this case is in fact annihilation and not real interpenetration. Thus, by default, extension is impenetrable. This impression is reinforced by his response to More's complaint that he cannot conceive of extended things moving in a plenum without some parts absorbing others.[8] To this, Descartes replies:

> I repeat here that if they are absorbed, then half the space is destroyed and ceases to be; but what ceases to be does not penetrate anything else; so impenetrability must be admitted in every space. (AT 5, p. 342; CSMK, p. 373)

More's description sounds as if this sort of absorption might be a quasi-natural process. Descartes's response fails to counter this impression and makes only the *ad hominem* point that even so, impenetrability should still be admitted, for annihilation is not penetration. But the impression is false. Descartes takes for granted that annihilation would require a special divine act, and that this is not something that actually occurs in the natural world. To put it another way, Descartes holds that when one body contacts another body, it must rebound or stop or push the other body;[9] that parts of it begin to disappear is not an option, absent a special (miraculous) act of God. That this is his view (that annihilation is not an option) is exhibited in the 6th replies, the only

[8] See Des Chene (*Physiologia*, p. 380) for a nice account of how More understands the possibility of this 'absorption', as condensation without loss of substance. Descartes, of course, held that this was impossible. In later work, More holds, as does Locke, that space is immobile and indivisible. See Reid, *Metaphysics of Henry More*.

[9] Of course, there are indefinitely many other options: it might crumble into tiny bits, flow around the body, etc.

place in the works published in his own lifetime in which he explicitly discusses the connection between impenetrability and extension.[10] That it *must* be his view is manifest throughout the *Principles*; notably, Descartes's claim in 2.33 that bodies can only move in 'circles' (closed curves) depends upon it:[11]

> I noted above that every place is full of bodies, and that the same portion of matter always takes up the same amount of space. It follows from this that each body can move only in a circle: a body entering a given place expels another, and the expelled body moves on and expels another, and so on, until the body at the end of the sequence enters the place left by the first body at the precise moment when the first body is leaving it. (P 2.33; AT 8A, p. 58; CSM 1, pp. 237–8)

For if bodies could, as it were, march into annihilation, then they could move in straight lines.

It is worth noting here that the very idea of the annihilation of extension is somewhat fraught, for Descartes. Could God annihilate a body by a special act? Surely Descartes would be loath to deny this. But there is a way in which such an act could not succeed, since the existence of any extension entails the existence of all of 'indefinite' extension, as follows from *Principles* 2.21, where Descartes argues that extension must be indefinite:

> What is more we recognize that this world, that is, the whole universe of corporeal substance, has no limits to its extension. For no matter where we imagine the boundaries to be, there are always some indefinitely extended spaces beyond them, which we not only imagine but also perceive to be imaginable in a true fashion, that is, real. And it follows that these spaces contain corporeal substance which is indefinitely extended. (AT 8A, p. 52; CSM 1, p. 232)

This might leave us wondering why annihilation plays such an apparently prominent role in Descartes's rhetoric in the argument for impenetrability. A good way to put it is this: Descartes is asserting that when you try to conceive of the overlap of two bits of extension, all you manage to conceive of is annihilation[12] (rather than asserting that when apparent overlap happens, that's because actual annihilation is happening). This fits with how he introduces the second version of the argument, which explicitly talks of what we can conceive.

How should we understand Descartes's argument, then? Among the few scholars to have commented on this issue, opinions differ wildly, both as to interpretation and as to evaluation. Daniel Garber writes (*Descartes' Metaphysical Physics*, p. 147) as though the claim that 'what is annihilated does not penetrate anything' is central to the argument, which, as I've just argued, it ought not to be. He also suggests a surprisingly rosy view of the argument's merits, calling it 'simple and ingenious'.

[10] 'For the true extension of body is such as to exclude any interpenetration of the parts' (AT 7, p, 442: CSM 2, p. 298). In *The World*, Descartes describes matter (prior to division) as a uniform, solid body filling three-dimensional space (AT 11, p. 33: CSM 1, p. 91).

[11] The rules of impact also presuppose it, of course.

[12] Thanks to Aaron Bronfman for a comment that suggested something like this formulation.

Alan Gabbey claims, in a typically rich footnote, that the coupling of impenetrability with extension 'is a logical consequence of the principle of the permanence of extended corporeal substance, and of the axiom *ex nihilo nihil fit*'.[13] I agree that we need the principle of the permanence of substance to rule out annihilation, but it is far from obvious what comes here from '*ex nihilo nihil fit*'. Perhaps Gabbey is suggesting the following: Suppose two pieces of extension shift so as to overlap. Then it would seem that a gap has opened up. But the 'gap', of course, is extended, so it is a body, and, supposing the overlap is genuine, you have more body than you used to have, but that violates '*ex nihilo nihil fit*'. This is a nifty argument, but I don't think it reflects Descartes's actual reasoning. He never suggests that were there overlap, new body would come to be with an inadequate cause. Rather, he argues that no overlap is possible, which is why he moves to the conclusion that any apparent overlap would have to be annihilation. (We might wonder in passing whether this reconstructed Gabbey argument, even though it is not Descartes, is compelling enough to do the work that Descartes wanted done. I suppose the reply would simply have to be: the extra extension is not brought into being by nothing, but by motion, and who says that geometrical extension is not the sort of thing that can be brought into being by motion?)

Bernard Williams is closer to the interpretative truth when he describes Descartes as 'relying merely on the logical truth that two things cannot be in the same place at the same time' (*Descartes*, p. 229). But is Descartes entitled to rely on this principle and is it a logical truth? Interestingly, Williams's own argument seems to undermine the thought that it is a logical truth, where he notes that it is legitimate to think of two geometrical solids as occupying the same space, or parts of them doing so.[14] The implication he takes from this is that Descartes's notion of extension is not purely geometrical. But surely it is entirely unclear why he is entitled to more than the purely geometrical notion, since bodies are supposed to be simply extension.

Dennis Des Chene turns Williams's discussion into a thought experiment, in effect, and offers two replies on Descartes's behalf, that is, two defences of the claim that Cartesian bodies cannot overlap. Imagine two overlapping polyhedra:[15]

> But suppose that from each polyhedron the part that does not belong to the other is truncated. What we have left is two polyhedra whose internal places coincide. But why just two? Why not infinitely many? (Des Chene, *Physiologia*, p. 381)

> ...suppose we take them [the two polyhedra] to be bodies, and consider the larger body that would result from joining them. That body would have at least two real—and not just imagined—parts that were not outside one another, and that could not subsist separately.
> (Des Chene, *Physiologia*, p. 381)

[13] Gabbey, 'Force and Inertia', p. 299, footnote 27.

[14] As Funkenstein notes (*Theology and the Scientific Imagination*, p. 74), this seems legitimized by Euclid, wherein geometrical objects pass through each other, coincide, etc.

[15] The example, borrowed from Williams, is of two polyhedra constructed on the same base.

Des Chene himself concludes that these inventive and illuminating replies do not settle the debate. The first, he briefly suggests, 'proves too much'. I think it might be more helpful to reply, 'why not indeed?' There are as many polyhedra there as there are, and whether or not we can discern whether there are one, two, or infinitely many is beside the point.[16]

The second is more revealing. It sounds definitive: The merged body would violate the standard characterization of extension in terms of *partes extra partes*. But would it? Descartes's imagined opponent holds that a duplicated interior is impossible for bodies, on account of solidity, but that it is conceivable for Descartes's merely geometrical objects. The place to push back is on the contention that these two internal parts could not subsist separately. Why not? To get the conclusion that they could not subsist separately, we must be presupposing what is at issue, that extended things, as Des Chene puts it, 'just have that mode of existence which entails that one [extended thing] is really distinct from another only if their locations are distinct'.[17] This is really another way of saying that two extended things cannot be in the same place at the same time. Thus, without begging this question, we cannot get inseparability.

Does this scenario violate '*partes extra partes*'? On the standard scholastic understanding of extension or extensive quantity, signalled by the phrase '*partes extra partes*', the parts in question must be both distinct in being and distinct in place (Des Chene, *Physiologia*, pp. 100–1). As just argued, nothing short of begging the question prevents us from asserting that all the parts of the merged overlapped body are distinct in being. The merged overlapped body would have parts that aren't distinct in place. But all this tells us is that the standard scholastic view rules out overlap, which fits with the fact that the standard scholastic view is that actually extended things are impenetrable. And this is all to point out that Descartes's views here are in many ways just good scholastic doctrine, which helps to explain why he thinks the move from extension to impenetrability is so unproblematic that it is unargued in the *Principles*.[18] However, if the question at issue is whether overlap is possible, it begs that question to simply assert that it is standardly assumed that the parts of extended things must be distinct in place.[19]

Interestingly, Robert Pasnau offers two quite different accounts of this argument. In a 2007 paper, Pasnau ends up concluding that the argument is circular. Indeed, he puts the central point quite nicely—a point that Bennett had earlier made against Descartes and Garber (Bennett, *Learning from Six Philosophers*, V. 1 p. 31). Consider two overlapping spheres.

[16] Although it seems that Locke himself may accept a version of this unpersuasive argument. See E II. xxvii. 2.

[17] Des Chene (*Physiologia*, p. 381) has 'body' instead of 'extended thing', but this seems potentially misleading, since Descartes's opponents will hold that a body is more than an extended thing.

[18] As other scholars have noted, e.g. Des Chene. Arguably Descartes's problems are created by the fact that he isn't *simply* a scholastic here, but is rather combining scholastic doctrine with a geometrical understanding of extension, bodies, etc.

[19] An obvious counter-proposal here is that you get extension as long as a thing has some parts that are distinct in place; it isn't required that all the parts be distinct in place. As is discussed in Section 4a, Locke waves off the Latin phrase.

The proponent of overlap will simply insist that in such cases the total volume of the overlapping objects does exceed the volume of the space in which those objects are located.

(Pasnau, 'Mind and Extension,' p. 302)

Thus to say, as Garber does (*Descartes' Metaphysical Physics*, p. 147), that 'If they do interpenetrate, then the total volume of A and B is less than the total volume of the two spheres before interpenetration' is just to beg the question—the defender of overlap will maintain that it is the same, not less, and that no reason against that claim has been presented.

In his recent book (*Metaphysical Themes*, pp. 319–21), Pasnau characterizes Descartes's argument as uniquely hard to evaluate and offers a diagnosis:

The reason the argument is so hard to evaluate, I have come to believe, is that it can be construed in two different ways: either as making a purely conceptual point about extension, or as making a claim about how bodies must behave in the natural world. (Pasnau, *Metaphysical Themes*, p. 319)

In effect, Pasnau here attempts to clear Descartes of charges of question-begging by taking the second tack, and transforming impenetrability into a law of nature, decreed and upheld by God. God has decreed: no overlapping matter, no 'telescoping universe'. This is an ingenious suggestion, but I think it is quite clear that it is not Descartes. For Descartes tells us that impenetrability is a property, like risibility in humans, derived from the essence. But on Pasnau's interpretation, impenetrability derives from God's *policies*, not from the essence of body, and would have a standing similar to a body's tendency to rectilinear motion.[20]

With Des Chene, I must conclude in the end that Descartes has begged the important question, but I don't think, as Pasnau seems to fear, that that is to accuse Descartes of sophistry. His arguments help to articulate a powerful intuition about spatiality, but one that seems consistently deniable.

3. Locke contra Descartes on Impenetrability

In examining Descartes's argument for the impenetrability of extension, we have been following an imagined reply to that argument, based on Bernard Williams's comment. But how would Locke, one of Descartes's actual opponents, reply? Garber, in a footnote,[21] suggests that Locke is among those who have missed Descartes's commitment

[20] Pasnau is aware of this. He explicitly has impenetrability rest on 'a kind of law-governed occasionalism' (*Metaphysical Themes*, p. 322). The problem, though, is that there is direct evidence that impenetrability is supposed to follow from extension: impenetrability is like risibility. Also, the arguments to More are surely supposed to show *that* and *how* impenetrability *follows*, not to explain how a law of nature applies here. Pasnau remarks that 'Descartes scholars seem generally to take for granted that impenetrability is supposed to follow from the nature of extension' (*Metaphysical Themes*, p. 320, footnote 17), but surely there is good reason why they do so. Pasnau also suggests that Garber may be happy with Descartes's argument because he has Pasnau's interpretation in mind. But Garber states (*Descartes' Metaphysical Physics*, p. 146) that impenetrability is a 'direct consequence of the very notion of extension'.

[21] Garber, *Descartes' Metaphysical Physics*, p. 148, p. 344 (endnote 105).

to impenetrability and thus have seen him, unjustifiedly, as simply conflating body and space. I think this greatly underestimates Locke.

It does seem arguably to be true of Leibniz, however. As Gabbey notes ('Force and Inertia,' p. 299) there is evidence that Leibniz ignored Descartes's commitment to impenetrability, since in his 1692 'Critical Thoughts on the General Part of the Principles of Descartes' he undertakes to lecture Descartes on the need to distinguish between hardness and impenetrability.[22] Descartes, of course, is well aware of the distinction, and agrees with Leibniz and Locke that while hardness should be relegated to the senses, impenetrability must belong to body as such. In addition (as Garber notes), Leibniz writes that Cartesians make bodies out of 'bare extension only' and neglect 'antitypy', that which renders bodies impenetrable to each other.[23]

Locke, I will argue, is not in this camp (if indeed the 'camp' extends beyond Leibniz). Indeed, if we return to E II. xiii, we can now see it as engaging directly and effectively with the Descartes of the correspondence with More.[24] This is hardly surprising, since the correspondence was well known and moreover is found in Locke's own library.[25] Contra Garber, although the charge of conflation is to be found in Locke, it is not because he has misunderstood Descartes. He does think Descartes has conflated distinct ideas, for he holds that the idea of solidity is included in the idea of body, but that, contra Descartes, it is *not* included in the idea of extension. Recall, though, that Locke does not simply accuse Descartes of conflating body with space, but specifically counters the claim from the correspondence with More that extension brings with it impenetrability or solidity by contending that the idea of space 'includes not the idea of solidity in it'.

Even more tellingly, his next move is to cut Descartes's argument off at its start; there is no separability in extension, so no transferability even 'in imagination'. This move might seem to be directed against the sufficiency of Cartesian physics (that is, you can't get a world of moving bodies by carving up geometrical extension), and it is that, but we should also see it as cutting off the argument from impenetrability. So understood, Locke is arguing thus: it may be that we cannot imagine parts of space penetrating each other, but that's because we cannot conceive of parts of space as movable at all.

It may seem, then, that we have traced this dispute back to an unsatisfyingly unsettleable disagreement about what is and is not conceivable. Interestingly, however, evidence is available that Descartes should have had some sympathy to Locke's intuition here, that extension without impenetrability is conceivable. In explaining to More why body cannot be defined by impenetrability, Descartes writes:

[22] Leibniz, *Philosophical Papers*, p. 392.

[23] In 'Conversation of Philarete and Ariste', Leibniz, *Philosophical Essays*, p. 258.

[24] Thus I would also counter Garber's characterization (*Descartes' Metaphysical Physics*, p. 148) of the correspondence as generally neglected.

[25] The correspondence is included in More's *Collection of several philosophical writings* (1662), which was a part of Locke's library as catalogued by Harrison and Laslett, LL 192.

... tangibility and impenetrability involve a reference to parts and presuppose the concept of division or limitation; whereas we can conceive a continuous body of indeterminate size, or an indefinite body in which there is nothing to consider except extension.

(P 2.33; AT 8A, p. 58; CSM 1, pp. 237–8)

Thus Descartes indicates that prior to division, impenetrability doesn't apply to extension. Locke's claim then is that to consider extension prior to division is to consider space. The idea we thus form is distinct from an idea of solid extension or body. Thus, Locke drives home his central point that we have distinct ideas of space (mere extension) and body. He claims further that having isolated the idea of extension, we should see that genuine divisibility in fact has no purchase here.

Have we caught Descartes in an inconsistency here? It may seem so: In arguing that extension is necessarily impenetrable, he had been arguing that extension implies impenetrability. But under pressure from More, he seems also to have said that extension isn't always impenetrable. That would be a contradiction. The key question is: Does Descartes mean to imply that actual division is required for impenetrability?[26] That is something he ought not to say. For one thing, it would be very peculiar; we might want to ask: Wasn't that stuff impenetrable before we divided it up, really? For another, it has as a consequence that extension isn't always impenetrable. I think the most charitable reading of Descartes here is something like this: It takes two steps to link impenetrability to extension.[27] We do so by way of division, which is, as it were, the middle term. We can consider indefinite extension without considering division. But if we do consider it we see that extension entails divisibility, which entails impenetrability.[28] Even on this most charitable interpretation, however, Descartes has come perilously close to acknowledging a key Lockean claim, about the availability of two ideas.

4. Objections (and Replies) to Locke's Position

Thus far, I have argued that viewed properly, Locke's critique of Descartes is appropriately targeted and, as far as we have seen, effective. I next want to consider a series of objections to Locke's argument against Descartes in E II. xiii. 11–14. I will defend him against the first two. The third objection is deeper, and I will suggest that it motivates and explains much of the rest of Locke's apparently disjoint remarks in E II. xiii. 15–20, as well as reflections that go beyond E II. xiii.

a. Partes extra Partes

One question provoked by the discussion above is this: Is Locke in violation of the standard characterization of extension as having *partes extra partes*? A full answer to

[26] Both Des Chene (*Physiologia*, p. 380) and Garber (*Descartes' Metaphysical Physics*, p. 148) seem to suggest that the answer is 'yes'.

[27] Thanks to Aaron Bronfman for comments that pointed me towards this formulation.

[28] Although impenetrability involves a reference to parts, on this interpretation, possible parts will do.

this question would require more space and expertise than I have available.[29] I think the short answer to this question is 'yes', but that Locke doesn't think so (here he is probably wrong) and doesn't care (here he is justified). We can see that Locke doesn't think that he's violated the *partes extra partes* doctrine from his cranky remarks in E II. xiii. 15:

If any one asks me, *What* this *Space*, I speak of, *is*? I will tell him, when he tells me what his *Extension* is. For to say, as is usually done, That Extension is to have *partes extra partes*, is to say only, That *Extension* is *Extension*: For what am I the better informed in the nature of *Extension*, when I am told, That *Extension is to have parts that are extended, exterior to parts that are extended*, i.e. *Extension* consists of extended Parts? (E II. xiii. 15)

As Woolhouse ('Locke's Idea of Spatial Extension', p. 315) nicely puts it, Locke sees '*partes extra partes*' as just a synonym or translation of the word 'extension', rather than as a definition. Further, he holds that what 'extension' stands for is simple, and thus not subject to definition.[30] Nevertheless, it does seem that Locke has violated the standard understanding of the phrase by maintaining that the parts of space are inseparable. For it was standardly held that, since the parts of extension are '*extra*' to one another, they are separable, that is, capable of separate existence.[31] Descartes, in attacking Henry More on divine extension, couples *parts without parts* with divisibility, and More fails to decouple them in his reply.[32] Of course, Locke has no interest in whether or not he's in accord with scholastic doctrine,[33] and we ought not to expect him to concede one of the central points at issue based on a stock understanding of an oft-used phrase. Furthermore, this becomes a much-disputed topic in the early modern period, and Locke is in good company on his side of the dispute, including the later More and Isaac Newton.

b. *Varieties of Separability*

This raises a related and more deeply philosophical worry, however, about Locke's treatment of separability. In arguing that the parts of extension are inseparable, Locke seems to be trying to focus on their moveability: can one part be removed (actually or mentally) from its neighbour, such that a distance now appears between them. But this, perhaps, makes separability too hard to get; it seems that what he should be

[29] Pasnau reports ('Mind and Extension', p. 306) that he has been unable to determine where this formula '*partes extra partes*' came from. He characterizes it as widespread in the medieval period (it is in Ockham, for example), but as absent in Aquinas or Augustine.

[30] Whether the idea in question really meets Locke's criteria for being simple is a tricky issue, as Locke reluctantly acknowledges in replying to Barbeyrac (E II. xv. 9, 5th edn.).

[31] See Des Chene, *Physiologia*, 99–101. Extensive quantity confers extension (being spatially distributed), divisibility, impenetrability. Suarez holds that if one part is outside another, both in being and in place, you have extensive quantity.

[32] See AT 5, pp. 274, 305; CSMK, p. 364. All that More affirms, here, however, is that since God doesn't have parts outside of parts, he isn't divisible. This doesn't commit him to divisibility following from *partes extra partes*. In later works, More will attribute parts outside of parts to God and spiritual substances, while denying the separability of the parts. See Reid, *The Metaphysics of Henry More*.

[33] In a journal entry from June 1676, Locke asserts that 'their partes extra partes' is 'proper only to body because body alone has partes and is divisible' but doesn't apply to distance (AG p. 77).

considering is whether the parts of space are really distinct or not, that is, whether one part of space may continue to exist even if it's neighbour is annihilated by God. One might even worry that Locke is conflating something like natural divisibility (which would entail movability) with metaphysical divisibility.[34]

While it is true that Locke is neglecting a salient way of understanding separability in this passage, there are nevertheless several things that can be said in his defence. I think Locke would also deny that we can conceive of God *annihilating* a part of space, such that there would then be 'superficies' where before there was continuity. Again, he would have good company here, including Newton.[35] Even more interestingly, that company arguably includes Descartes himself, since any 'hole' purportedly created by annihilation would be extended and thus would be body, just as before. More broadly, as pointed out in Section 3, for Descartes, the existence of the whole of indefinite extension is implied by the existence of any part of extension. What Descartes clearly does think possible, of course, is for the parts of extension to be actually divided by motion (P 2.23; AT 8A, pp. 52–3; CSM 1, p. 232). Thus, Locke's emphasis on moving or removing as the means of separation is perfectly Cartesian; his response is appropriately targeted.

c. *Ontology (E II. xiii. 15–20) and Traditional Metaphysical Categories*

A third objection, and the central objection anticipated by Locke, is a metaphysical one. Locke expects the Cartesian to ask, 'What is the ontological status of this space?' and to flaunt her own purportedly straightforward answer to that question.[36] Locke's response, including as it does the infamous passage about the 'intelligent American' (and what he would say if instructed that a basis is what supports a pillar and a pillar is what is supported by a basis) is easily dismissed as an extended rant against the notion of substance, exactly the sort of thing that (understandably) misled Stillingfleet into thinking that Locke had discarded substance out of the world (LW 4, p. 5). It is important to notice, however, that Locke's response is in fact structured and, again, carefully targeted.

The first question that Locke attributes to his metaphysically minded Cartesian (in E II. xiii. 15) is 'What is space?', to which, as we've already seen, Locke replies (in effect), 'whatever extension is'. Of course, as Locke well knows, he must dissent from the full Cartesian answer to the question, 'what is extension?', since for Locke it certainly isn't material substance. The next move he gives to his Cartesian interlocutor brings this disagreement to the fore.

Those who contend that *Space and Body* are *the same*, bring this *Dilemma*. Either this *Space* is something or nothing; if nothing be between two Bodies, they must necessarily touch; if it be allowed to be something, they ask, whether it be Body or Spirit? To which I answer by another Question, Who told them, that there was, or could be nothing, but solid Beings, which could not think; and thinking Beings that were not extended? (E II. xiii. 16)

[34] I am influenced here by Holden, *Architecture of Matter*, pp. 11–16.

[35] See e.g. Newton in 'De Gravitatione', *Philosophical Writings*, 25–6.

[36] 'Why, it is extension, the attribute of all corporeal substance!' would be the purportedly straightforward answer.

Of course, as Locke sees it, this is a false dilemma. The first part of the argument he sees as a bad play on words, easily answered by observing that pure space may lie between the two bodies. The second is a bit of question-begging in favour of strict Cartesian dualism. It is easy enough to see where he finds the first part of the argument—Descartes himself reports that he has 'often said, nothingness cannot possess any extension' (P 2.18; AT 8A, p. 50; CSM 1, p. 231; see also the correspondence with More, discussed below in this section). It is less obvious where one might find Descartes giving part two of the argument. It seems rather that the direction of Descartes's argument is this one: If extension lies between the two bodies, then extended substance must lie there. But extension is the essence of body, all that is required for body, so body lies there. If one baulks in a Lockean way and insists that solidity too is required, Descartes can bring in the (disputed) argument for impenetrability to insist that all extended things are impenetrable and so obviously corporeal.

I suggest, however, that it is not difficult to read section 9 of Part 2 of the *Principles* as the sort of argument from dualism that Locke attacks here:

> 9. *If corporeal substance is distinguished from its quantity, it is conceived in a confused manner as something incorporeal.*
>
> Others may disagree, but I do not think they have any alternative perception of the matter. When they make a distinction between substance and extension or quantity, either they do not understand anything by the term 'substance', or else they simply have a confused idea of incorporeal substance, which they falsely attach to corporeal substance; and they relegate the true idea of corporeal substance to the category of extension, which, however, they term an accident. (P 2.9; AT 8A, p. 45; CSM 1, pp. 226–7)

Now, Descartes's real target here is a scholastic distinction, and what is directly impugned is the attempt to conceive of body independently of quantity. Nevertheless, in the background is exactly what Locke objects to, namely, the tendentious Cartesian view that there are only two natures or principal attributes available to be grasped, and they are extension and thought.

In sections 17–20, Locke broadens his attack into an assault on traditional metaphysical categories of substance and accident:

> If it be demanded (as usually it is) whether this *Space* void of *Body*, be *Substance* or *Accident*, I shall readily answer, I know not: nor shall be ashamed to own my Ignorance, till they that ask, shew me a clear distinct *Idea* of *Substance*. (E II. xiii. 17)
>
> But were the Latin words *Inhærentia* and *Substantia*, put into the plain English ones that answer them, and were called *Sticking on*, and *Under-propping*, they would better discover to us the very great clearness there is in the Doctrine of *Substance and Accidents*, and shew of what use they are in deciding of Questions in Philosophy. (E II. xiii. 20)

Though by the end Locke is perhaps ranting more than arguing, there are at least two serious philosophical points to be found here. The first is that our notions of substance and accident are not sufficiently clear and explanatory enough for us to use them as weapons in philosophical argument. Although Locke's denunciations here are emphatic enough to have misled Stillingfleet, they are in fact consistent with his considered beliefs that (1) the world is full of substances and (2) we need to use a (modest) idea of substance to think about the world.[37] So, one response he has to the question of the ontological status of space is to wave it off as not worth addressing unless it can be more clearly formulated. A second point is that if we are willing to call body, spirit, and God all substances, nothing prevents us from regarding space as a fourth kind of substance. 'And if they can thus make three distinct *Ideas* of *Substance*, what hinders, why another may not make a fourth?' (E II. xiii. 18). That space fits poorly into traditional ontological categories, but is none the worse for that, is a view that Locke shares with Newton and Gassendi. As applied to Descartes, the complaint is again that Cartesian dualism is unjustifiedly strict.

We might ask, also, whether an extended rant about the overconfident use of substance and accident can reasonably be seen as provoked by Descartes. We have already seen in the *Principles* Descartes's readiness to argue that extension cannot belong to nothing but must be attributed to a substance (P 2.16; AT 8A, p. 49; CSM 1, p. 229). Here again, we should see the correspondence with More as also in the background. Descartes's discussion of impenetrability is part of an effort to suggest to More that the attribute of extension is so weighty, as it were, that it could only be supported by corporeal substance. Descartes follows his description of the transferability of the parts of extension and their impenetrability with this appeal:

> Now since I consider that such real properties can exist only in a real body, I dared to assert that there can be no completely empty space, and that every extended being is a genuine body.
>
> (AT 5, p. 271; CSMK, p. 362)

Descartes thus seems to be arguing that 'all the properties of extension' must be supported by some substance, and, indeed, a substance of the right kind, body. (God, as he argues just below, is a substance of the wrong kind.) Although it is perhaps uncharitable of Locke to lump Descartes together with an unnuanced version of scholasticism, it is not unreasonable of Locke to view Descartes here as engaging in the sort of explanatory use of 'substance' to which Locke objects.

Locke's own response to the ontological challenge in E II. xiii is, as we have seen, non-committal. He indicates here that he sees no barrier to regarding space as a kind of substance, different from body, spirit, and God. He further insinuates that the ontological challenge ought not to be given much argumentative weight, at least not if framed in traditional categories of substance and accident. It would be premature, of course, to conclude that Locke has no ontology of space. That is a topic that I will defer,[38] though it is worth observing that E II. xv. 3 betrays yet more affinities between Locke and More (and Newton), as Locke suggests explaining space in terms of the way in which God 'fills immensity'.

[37] See McCann, 'Locke on Substance', and Downing, 'Sensible Object and Uncertain Philosophical Cause'.

[38] See Downing, 'Locke and Descartes', pp. 115–16.

5. Conclusion

What I hope to have shown, then, is that these few sections of E II. xiii display a deep critical engagement between Locke and Descartes on the metaphysics of body, one which is best understood in relation to the earlier profound dispute between Descartes and More. I have argued that Locke is a more astute critic of Descartes than he is sometimes thought to be, and a more thorough and subtle reader of Descartes than he himself is willing to admit.

I see this as part of a broader case that Locke, rather than rejecting Descartes's philosophy for methodological reasons, qua empiricist, as is sometimes thought, in fact engages with the details of his metaphysical positions. Locke's objections to Cartesian ontology are not to the very doing of ontology, and little to the way in which Descartes does ontology,[39] but rather to the arguments by which the ontology is supported and to the positions that result. Indeed, far from rejecting Cartesian methodology, Locke seems to sarcastically invoke it against Descartes himself, in concluding his official consideration of space:[40]

> 'Tis not easie for the Mind to put off those confused Notions and Prejudices it has imbibed from Custom, Inadvertency, and common Conversation: it requires pains and assiduity to examine its *Ideas*, till it resolves them into those clear and distinct simple ones, out of which they are compounded; and to see which, amongst its simple ones, have or have not a necessary connexion and dependence one upon another: Till a Man doth this in the primary and original Notions of Things, he builds upon floating and uncertain Principles, and will often find himself at a loss. (E II. xiii. 27)

References

Bennett, Jonathan. *Learning from Six Philosophers*. 2 vols.V ol. 1. Oxford: Oxford University Press, 2001.

Des Chene, Dennis. *Physiologia*. Ithaca, NY: Cornell University Press, 1996.

Downing, Lisa. 'The "Sensible Object" and the "Uncertain Philosophical Cause"', in Daniel Garber and Béatrice Longuenesse, eds., *Kant and the Early Moderns*. Princeton: Princeton University Press, 2008: 100–16.

Downing, Lisa. 'Locke and Descartes', in Matthew Stuart, ed., *A Companion to Locke*. Oxford: Blackwell, 2015: 100–20.

[39] This is not to say that Locke has no systematic objections to Cartesian metaphysics. Clearly he sees Descartes as typically dogmatic and as too ready to poach on empirical preserves, e.g. foreclosing the possibility of a vacuum, or of a mind ceasing to think, by stipulation.

[40] This chapter has been presented as a talk over several years, and so I have accumulated many debts (some of which I may have forgotten). I have benefited greatly from audiences at the University of Colorado, the University of Nebraska, University of California Irvine, University of California Berkeley, University of Wisconsin Madison, the University of Missouri, Simon Frasier University, Illinois State University, and the University of Virginia, as well as the 8th Nordic Early Modern Philosophy Workshop. Many thanks also to Philippe Hamou and two anonymous editors for helpful comments that improved the chapter.

Funkenstein, Amos. *Theology and the Scientific Imagination: From the Middle Ages to the Seventeenth Century*. Princeton: Princeton University Press, 1986.

Gabbey, Alan. 'Force and Inertia in the Seventeenth Century: Descartes and Newton', in Stephen Gaukroger, ed., *Descartes: Philosophy, Mathematics, and Physics*. Hassocks: Harvester Press, 1980: 230–320.

Garber, Daniel. *Descartes' Metaphysical Physics*. Chicago: University of Chicago Press, 1992.

Holden, Thomas. *The Architecture of Matter: Galileo to Kant*. Oxford: Clarendon Press, 2004.

Leibniz, G. W. *Philosophical Papers and Letters*, ed. and trans. Leroy E. Loemker. The New Synthese Historical Library. 2nd edn. Dordrecht: Kluwer, 1969.

Leibniz, G. W. *Philosophical Essays*, ed. and trans. Roger Ariew and Daniel Garber. Indianapolis: Hackett, 1989.

McCann, Edwin. 'Locke on Substance', in Lex Newman, ed., *The Cambridge Companion to Locke's 'Essay Concerning Human Understanding'*. Cambridge: Cambridge University Press, 2007: 157–91.

Newton, Isaac. *Philosophical Writings*, ed. Andrew Janiak. New York: Cambridge University Press, 2004.

Pasnau, Robert. 'Mind and Extension (Descartes, Hobbes, More)', in *Forming the Mind: Essays on the Internal Senses and the Mind/Body Problem from Avicenna to the Medical Enlightenment*. Dordrecht: Springer, 2007: 283–310.

Pasnau, Robert. *Metaphysical Themes* 1274–1671. Oxford: Oxford University Press, 2011.

Reid, Jasper. *The Metaphysics of Henry More*. International Achives of the History of Ideas. Dordrecht: Springer, 2012.

Williams, Bernard. *Descartes: The Project of Pure Enquiry*. Hassocks: The Harvester Press, 1978.

Woolhouse, R. S. 'Locke's Idea of Spatial Extension', *Journal of the History of Philosophy*, 8/3 (1970), 313–18.

5

Modes and Composite Material Things According to Descartes and Locke

Martha Brandt Bolton

With regard to ontology, Descartes and Locke can be classed as post-scholastic substance theorists. Each posits exactly two types of entities: substances and things which depend for their existence on substances, the latter variously called 'accidents', 'affections', 'attributes', 'qualities', or 'modes'. In the abstract, this scheme can be compared to Aristotle's doctrine in *Categories*. Aristotle argues that some things exist in another; some are said of another; and some neither exist in nor are said of anything. Things of the first two sorts, which are called 'accidents', depend for their existence on things of the last sort, which are thereby shown to have the fundamental status of substances, as Aristotle argues; for example, Bucephalus is a substance. Descartes and Locke hold theories which can be traced to this source. But they are broadly speaking mechanists who hold that matter merely as such is a substance and the nature of material substance is homogeneous; material substance is not divided by species, or kinds. As a result, nothing classed as a substance by the Aristotelians is considered to be strictly speaking a substance by the two early moderns. For Aristotelians, every substance is a union of matter with species determining form; Bucephalus is thus constituted of appropriate matter and equine form. For the two early moderns, nothing is a union of matter and form; matter is a substance and although Bucephlus is material he is not a substance.

Descartes's commitment to mechanism is dogmatic, whereas Locke's is tentative and hypothetical. According to Descartes, the essence of material substance is the attribute extension, i.e. geometrical extension or what scholastics called 'continuous quantity'. According to Locke, extension and solidity are universal and basic qualities of corporeal substance insofar as we conceive it. They agree that every extended thing involves actual extended parts; this is not to say that the parts are, in fact, ever separated but just that this is possible in principle. Now the extended parts of an extended substance might, or might not, be regarded as so many distinct substances. Descartes does regard them this way. Locke posits atoms with naturally inseparable parts which are not explicitly said to be substances, but compose atoms each of which is a substance numerically distinct from every other. Extended substances are diversified by entities in the category accidents, qualities, or modes. For the mechanists, there are relatively few such entities: figure, size,[1] motion or rest, position, and texture. All other qualities, dispositions, and operations of bodies are supposed to be reducible to various combinations of determinates of these qualities. A thing which is constituted by entities in both categories cannot be identified with its substantial component. Bucephalus is not a substance in which certain qualities inhere. He is constituted as an entity by modes and several substances of exactly the same kind as those found in all material things. What makes him a horse are the modes.

Nevertheless horses, water, and other sensible material things are substance-like; they have the metaphysical properties which are distinctive of Aristotelian substances. Water and gold are subjects of which things are predicated, subjects in which qualities are combined in a recurrent and explicable way. In addition, we tend to suppose that a horse initiates change, undergoes change, and exists at different times. Aristotelians take these patterns of unity to be marks of a substance, whereas the two early moderns are committed to the theory that they are marks of a multitude of substances and modes. To the extent the moderns purport to account for the metaphysical properties of sensible things, such as being one thing to which several qualities are ascribed, being individually distinct, and remaining the same over time, they must do so largely on the basis of modes. This is the general thesis of the present chapter. Descartes and Locke share two general anti-Aristotelian views. (1) There are many individually distinct material substances on which affections, qualities, and modes depend. (2) A sensible thing such as Bucephalus is not properly regarded as a substance, but rather as a composite of substances and modes in which modes individuate and unify the composition. The result is a post-Aristotelian reconfiguration of matter, substance, modes, and sensible material things.

1. Descartes on Modes, Extended Substances and Mechanist Bodies

In some texts, Descartes invokes modes to explain the notion of substance: a substance is anything 'in which whatever we perceive immediately resides, as in a subject' ('Second Set of Replies'; AT 7, p. 161: CSM 2, p. 114). According to a different text, a substance is a thing whose existence depends on nothing else and there is only one substance, God; but created substances are things whose existence depends on nothing

[1] As we will see, for Descartes, size understood as a magnitude may be a mode, but what is measured by the magnitude, i.e. a quantity of extension, is not. Size is mentioned as a 'corporeal act' along with figure, motion, and 'others which cannot be thought without local extension', that is to say, among modes at 'Third Objections and Replies', reply to Second Objection, AT 7, p. 176; CSM 2, p. 124. Size is listed as an attribute in a letter to *** 1645 or 1646, AT 4, p. 349: CSMK, p. 280.

but the concurrence of God.[2] In this context, the things we immediately perceive, that is modes, are characterized as depending on created substances in a way in which no substance depends on anything.[3]

Principles 1.56 clarifies the difference among attributes, qualities, and modes. A dependent entity is a mode if it modifies, affects, or alters the thing in which it resides. A mode is a quality if it shows that the thing in which it exists belongs to a certain kind.[4] Finally, entities which depend on a substance and remain unchanged should be called 'attributes'. Every substance consists of the attributes existence, duration, and number; every created substance has in addition a principal attribute which is its essence—extension, in the case of corporeal substance, and thought, in the case of a mind. But although each attribute of a substance has a certain dependence on that substance, the relation is symmetrical; in the existing thing, the substance is no different from its attributes.[5] None of them can be clearly and distinctly perceived without the others, although each can be partially conceived without the others by a mental act of abstracting (ignoring) them.[6] Attributes, then, are just substances differently (incompletely) conceived and qualities are just modes by which we identify various kinds of things. Hence, 'We recognize no other categories of things apart from substances and their modes.'[7]

The dependence relation between modes and substances is characterized epistemically in the Cartesian texts. As we saw, modes are the entities we immediately perceive by which we come to know the existence and nature of substances.[8] According to *Principles*, the distinction between a mode and the substance to which it belongs can be recognized by a conceptual test: 'we can clearly perceive a substance apart from the mode which we say differs from it, whereas we cannot, conversely, understand the mode apart from the substance' (P 1.61; AT 8A, p. 29; CSM 1, p. 213). A spherical shape, e.g., cannot be grasped apart from an extended thing. To be more exact, the concept of the figure contains the concept of an extension: 'the nature of a mode is such

[2] *Principles of Philosophy* (hereafter P), 1.51; AT 8A, p. 24; CSM 1, p. 210.

[3] '...when we call a created substance self-subsistent, we do not rule out the divine concurrence, which it needs in order to subsist. We mean only that it is the kind of thing that can exist without any other created things; and this is something that cannot be said about the modes of thing, like shape and number.' Descartes to Hyperaspistes, August 1641; AT 3. 429: CSMK, pp. 193–4.

[4] P 1.56; AT 8A, p. 26; CSM 1, pp. 211–12.

[5] An existing substance and its nature are particular: '[B]y thought, I do not mean some universal which includes all modes of thinking, but a particular nature which takes on these modes.' (Descartes to [Arnauld], 29 July 1648, AT 5, p. 221; CSMK, p. 357). Chappell, 'Descartes's Ontology', pp. 111–13, maintains that the metaphysical scheme in *Principles* posits 'omni-generic' attributes, existence, duration, number, and 'uni-generic' attributes, thinking and extension, and says that the former can be conceived without the latter. This may be true of genera, but it is not true of the existence, duration, and thinking that exist in an individual substance; see P 1.62.

[6] See P 1.62. In general, things that are conceptually distinct are different in thought, but 'in no way distinct' in the thing existing outside of thought; see Descartes to ***, 1645 or 1646; AT 4, p. 349; CSMK, p. 280 and Descartes to Mesland, 2 May 1644; AT 4, p. 116: CSMK, p. 234. See Descartes to Mesland, 2 May 1644; AT 4, p. 116; CSMK, p. 234 on abstractive and exclusive conception.

[7] P 2.55; AT 8A, p. 71; CSM 1, p. 246.

[8] This method is pursued in detail in the *res cogitans* passage in Meditation 2.

that it cannot be understood at all unless the concept of the thing of which it is a mode is implied in its own concept.'[9] That a mode cannot be conceived apart from an attribute implies that the attribute partially defines the essence of the mode.[10] By contrast, the attribute can be conceived apart from the mode.[11] The attribute is conceived as a substantial nature and conceived as being (receiving, having) any of the modes it partly defines, but not conceived in this way as having any other modes.[12] Some scholars suggest that ontologically the attribute–mode relation is assimilated to the relation between a determinable and its determinants.[13]

This paves the way for the real distinction between thinking and extended substances. Descartes argues from the modes by which we come to know a substance: '[W]e do not come to know a substance immediately, through being aware of the substance itself; we come to know it only through its being the subject of certain acts.'[14] We cannot perceive a figure or motion without forming a concept which contains that of an extended thing and we are said to find that its being extended suffices for it to be a 'complete being', that is, a substance. For according to one definition, a substance is anything in which several modes (acts) reside as in a subject.[15] Applied to an act of thought, similar reasoning shows that thinking is enough to constitute a substance. Moreover, one nature is not a genus of which the other is a species. If that were the case, the concept of one would contain that of the other, and we do not find that this is true.[16] A thinking substance and an extended substance can, then, each be clearly and distinctly understood apart from the other. If each of two things can be distinctly understood apart, they can 'exist apart', as Descartes has it. This is the criterion for the real distinction which holds among all and only substances.[17] Expressed in terms of the proposed model: figures and motions (affirmations and volitions) are determinants of determinable extension (thought) but not determinable thought (extension); and extension and thought are highest determinables.

Yet the modes of extended substance depend not so much on the entire substance as on one or more of its parts. What modes reside in as in a subject, is not *res extensa* in its entirety, but local parts of it. It is in just these terms that Descartes explicates

[9] 'Comments on a Certain Broadsheet', AT 8B, p. 355; CSM 1, p. 301.

[10] The attribute is 'included in the essential definition' of the modes [*in suo formali conceptu includunt*]; Meditation 6, AT 7, p. 78: CSM 2, p. 54.

[11] P 1.61. [12] P 1.63–4; also see Sixth Set of Replies, AT 7, p. 423; CSM 2, p. 286.

[13] See e.g. Secada, *Cartesian Metaphysics*, pp. 184–96.

[14] Third Objections and Replies, reply to Third Objection; at AT 7, p. 176; CSM 2, p. 124.

[15] Second Set of Replies, AT 7, p. 161; CSM 2, p. 114; the idea of a mode has objective reality according to AT 7, p. 165.

[16] See especially Fourth Set of Replies, AT 7, pp. 221–2: CSM 2, pp. 156–7; Bolton, 'Thinking: The Nature of Descartes' Mental Substance', 65–81 argues for this interpretation in more detail. Rozemond, 'Descartes' Dualism', 372–89 proposes an argument with an unsupported premise to the effect that a substance has exactly one principle attribute. As I see it, Descartes needs to show that this is true, e.g. to counter Hobbes's assumption that a substance has several faculties each with a different essence. In *Meditations* and *Principles*, he purports to prove it on the basis of the conceptual connections mentioned in the text.

[17] P 1.60; AT 8A, pp. 28–9: CSM 1, p. 213; Meditation 6, AT 7, p. 78: CSM 2, p. 54.

each of the modes of extension: 'shape, which is a function of the boundaries of this extension; position, which is a relation between various items possessing shape; and motion, or change in position' (Meditation 3; AT 7, p. 43: CSM 2, p. 30). Boundaries mark limits of extension, positions presuppose limits, and motion presupposes position, so all modes have concepts which include limited extension. Indeed the logic of determinables and determinants shows that modes do not belong to the whole of extended substance. No two determinants of the same determinable can exist in the same thing at the same time. Spheres, cubes, moving things, and resting ones are present in corporeal substance at the same time, as we find in experience. We avoid contradiction by supposing that shapes and motions exist in different parts of extended substance.[18]

But it is difficult to say just which part of *res extensa* a given mode exists in because Descartes provides no objective account of what individuates parts of matter.[19] This is due to his liberality about numerically different extended substances. *Principles* 1. 60 explains that the real distinction holds between all and only substances. The criterion for things that are really distinct is that 'we can clearly understand one apart from the other'. The article goes on to say that, assuming extended substance exists, we can be certain that 'every part of it, as delimited by us in our thought, is really distinct from the other parts of the same substance'.[20] It is very clear that the number of things there are depends on this distinction, and the real distinction pertains to substances; accordingly, any arbitrary partition of extended substance yields entities each of which is a substance numerically distinct from the others.

This is the ontological ground of the divisibility of matter:

> ...we can say that the existence of atoms, or parts of matter which have extension and yet are indivisible, involves a contradiction, because it is impossible to have the idea of an extended thing without also having the idea of half of it, and so conceiving it as being divisible by two or three. From the simple fact that I consider the two halves of a part of matter, however small it may be, as two complete substances, whose ideas are not made inadequate by an abstraction of my intellect, I conclude with certainty that they are really divisible.
>
> (Descartes to Gibeuf, 19 January 1642; AT 3, p. 476: CSMK, pp. 202–3)

Extended substance is divisible along every closed boundary because every such boundary individuates a substance numerically distinct from the contiguous substance. The texts are explicit and the point is important to Cartesian physics. Still several scholars argue against the literal interpretation of the two passages just quoted. Perhaps the most persuasive argument is that the real distinction is known by intellect alone. For Descartes, there is no intrinsic difference between essences which is not thoroughly intelligible; if two parts of extended substance have essences which are in all intelligible intrinsic respects the same, as they would seem to be, they are not distinct

[18] For a different view, see Lennon, 'Descartes' Idealism', 53–6.

[19] See Lewis, *L'Individualité selon Descartes*, 39–51.

[20] P 1.60; AT 8A, pp. 28–9: CSM 1, p. 213.

Cartesian substances, or so the argument goes.[21] But although the difference between minds and bodies is intelligible, the claim that this is true of all substances, including those of the same kind, is not explicit in the text. The multiplicity of extended substances is explicit, and it can be supported by reasoning along a line laid out by the general theory of substance. Because any extension delimited in thought is constituted by nothing outside its boundary, its essential definition refers to nothing extrinsic to that; so it can be conceived apart from any portion of extension to which it is contiguous.[22]

But although Descartes is committed to the existence of many individually different corporeal substances, consistency requires that their numerical distinctness be conditional. If we mark off a limited portion of extension, we can equally well partition it into two or three or more individually different parts. The real distinction is the basis of enumeration of substances, and it is incoherent that one substance is many substances.[23] Moreover, if one extended substance contained several substances numerically distinct *from it*, the one would be partially coincident with the several others. It would be as if the parts were penetrated by the whole, which is impossible by Cartesian principles.[24] Descartes argues that if two extended substances were in the same place at the same time, there would be, in total, exactly as much extension as there would be if only one of them were there; that is to say, the extension of one of them would be annihilated and only one substance would be located in this place. Given that any way of dividing substances *is* a way of enumerating them, it is best to assume that any division is conditional on a way of thinking about extended stuff. This is only implicit, but consistency demands something to this effect. If we understand the multiplicity doctrine this way, all internally consistent specifications of boundaries are equally legitimate ways of individuating limited substances, but none of them is absolute.[25] If two schemes are accepted at once, inconsistencies follow. Still, relative to the delineation of any region of *res extensa* one chooses, it is a quantity of extension; there is only one quantity, and it cannot be other than a substance. For 'There is no real difference between quantity

[21] Secada, *Cartesian Metaphysics*, 209. Other scholars who maintain that Descartes posits only one extended substance include Williams, *Descartes: The Project of Pure Reason*, pp. 127–8); Keeling, *Descartes*, 280–3; Dicker, *Descartes*, 292–30, which takes a passage in the 'Synopsis' of the *Meditations* (AT 7, pp. 12–16: CSM 2, pp. 9–11) to count against the multiplicity of extended substances, whereas I take it to count against the indestructibility of non-substantial bodies; Lennon, 'Descartes' Idealism', 'The Problem of Individuation among the Cartesians'.

[22] This is not to deny that a limited extension exists only if unlimited extension exists outside its boundary; see Descartes to More, 15 April 1649; AT 5, p. 347: CSMK, p. 375. It is possible for a finite quantity of extension to exist apart from the entire infinite quantity that surrounds it, although if it were to do so, God would have to create some infinite quantity of extension around it; see Normore, 'Descartes and the Metaphysics of Extension', p. 283.

[23] On the theory that a whole is nothing but its parts, the difficulties about numeration and penetration are avoided. On this theory, a substance composed of, say, two substances would be two substances, not three. There would be nothing that coincides with the whole other than it. But as far as I can see, there is no explicit textual evidence that Descartes subscribes to this theory of wholes and parts and doing so would undermine the authority of the real distinction in questions of number.

[24] P 2.7–8.

[25] Compare Stuart, 'Descartes' Extended Substances', 82–104, which also maintains that Descartes is committed to what he says in P 1.60.

and the extended substance; the difference is merely a conceptual one' (P 2.8; AT 8A, p. 44: CSM 1, pp. 226–7).

Indeed, a given quantity remains the same in change: '... one and the same body, with its quantity unchanged, may be extended in many different ways (for example, at one moment it may be greater in length and smaller in breadth or depth, and a little later, by contrast, it may be greater in breadth and smaller in length)' (P 1.64; AT 8A, p. 31; CSM 1, p. 215). But '... it is impossible to take even the smallest fraction from the quantity or extension [of a body] without also removing just as much from the substance' (P 2.8; AT 8A, p. 45; CSM 1, p. 226). Suppose a quantity of extension is 'continuous' in the sense that between any two points on or within its boundary there is an uninterrupted path which lies entirely on or within its perimeter, and then suppose a part is spatially separated from the rest. What remains continuous is not the same quantity, but that original quantity cannot cease to exist. A substance is naturally indestructible, needing nothing but the concurrence of God. How a quantity might be traced if its continuity is disrupted is not explained in the texts; this is a lacuna in Descartes's account of the identity of a limited corporeal substance. Assuming that the identity of a substance is in question,[26] it seems Descartes wants to say there is no absolute answer to this question, but only answers relative to ways of individuating substances.

To put this differently, there is tension in the Cartesian assimilation of geometrical space and corporeal substance because mathematical space has a structure seemingly inconsistent with that of Cartesian extended substance.[27] Mathematical extension can be arbitrarily divided into regions some of which overlap others, but any partition of *res extensa* is a division of individual substances which are mutually impenetrable on Descartes's principles. Still he contrives to impart to substantial extension a structure which is not absolute but relative to ways of designating parts, which is equivalent to the structure of geometrical space in respect of the number of different parts it contains and the variety of figures. *Res extensa* can be conceived to have arbitrarily chosen parts; all parts are available, but some are incompatible with others. It can accordingly receive all compatible motions and figures precisely because it contains whatever individual substances might be needed to serve as local subjects in which actual modes reside.

In creating matter, God creates a certain distribution of motion and rest. With motion comes all the variety in matter, all the qualities which diversify material things according to mechanism:

All the properties which we clearly perceive in [matter] are reducible to its divisibility and consequent mobility in respect of its parts, and its resulting capacity to be affected in all the ways which we perceive as being derivable from the motion of its parts. If the division into parts

[26] As we shall see, Descartes's use of 'body' is ambiguous as between an extended substance and actual material things.

[27] Some scholars conclude that the assimilation is inconsistent, e.g. Keeling, *Descartes*, p. 80; Grosholz, 'Descartes and the Individuation of Physical Objects', 41–58; Dicker, *Descartes*, pp. 295–301.

occurs simply in our thought, there is no resulting change; any variation in matter or diversity in its many forms depends on motion. (P 2.23; 8A, pp. 52–3: CSM 1, p. 232)

Motions effect a separation of different parts of matter, local bits of matter which slide by each other along a common surface. As this passage suggests, a motion realizes a separation which is possible in virtue of the thought-relative individuation of substances comprised in *res extensa* independently of its modification by motion. We may suppose a motion resides in a subject consisting of one of that pair of extended substances whose mutual boundary in thought is precisely the surface along which mutual separation occurs.[28] The existence of the motion gives these particular substances a role in the mechanism of nature.

Mechanist theory purports to account for all qualities, or modes, of bodies on the basis of motions. I will call the bodies relevant to such explanations 'mechanist bodies'. In view of the primacy of motions in the explanation of all other bodily modes and the intrinsic homogeneity of extended substance, there is nothing by which mechanist bodies might be differentiated but their motions.[29] That motion alone differentiates bodies is clear from the explanation of 'motion in the strict sense' offered in *Principles*:

...motion is the transfer of one piece of matter, or one body, from the vicinity of the other bodies which are in immediate contact with it, and which are regarded as being at rest, to the vicinity of other bodies. By 'one body' or 'one piece of matter' I mean whatever is transferred at a given time, even though this may in fact consist of many parts which have different motions relative to each other. (P 2.25; AT 8A, pp. 53–4; CSM 1, p. 233)[30]

Some scholars think this is circular: 'one body' is defined in terms of motion (transfer), but the definition of motion takes the notion of one body for granted.[31] Yet the passage fully comports with what we have said about Descartes's ontology. It is best understood not as a definition, but as an account of how moving bodies are metaphysically constituted. In particular, a moving body is individuated by its motion. Motions are caused in accord with the laws of motion. Because a motion cannot be conceived apart from a substance and the moving body has a finite extension, the body also comprises an extended substance; namely, that limited substance whose boundary in thought aligns with the surface cut by the motion. To put it differently, a mechanist body is a motion and whatever substance(s) are required for it to exist. Contrary to what we might assume, a moving body is not an 'enmoded substance', [32] individuated as a substance and contingently modified by a certain motion. Instead it is an 'ensubstanced mode'; its being

[28] Strictly speaking, a motion exists in each of the separating bodies, but in order not to depart too far from ordinary ways of speaking, Descartes assigns motion to one of the bodies and considers the other to be at rest (P 2.29).

[29] Leibniz argues that motion, considered as mere transfer without reference to forces which cause it, is not sufficient to make a difference in the homogeneous extended plenum. But Leibniz's reasoning presupposes the principle of the identity of indiscernibles which is not shared by Descartes.

[30] Similarly, *Traité de Lumiere*, AT 11, p. 15. Also see P 2.28.

[31] See e.g. Garber, *Descartes' Metaphysical Physics*, 176–81.

[32] I take this phrase from Geoff Anders.

an individual body is due to its motion, and the substance it comprises is identified relative to that. The motion requires the thought-individuated substance that has the appropriate boundary. By attaching this substance, the motion incorporates it in a body and realizes it in the world.

A mechanist body is not a substance, but its constitution includes one or more extended substances. Taken together, the mechanist bodies that exist at the same time realize a proper subset of the collection of available extended substances. Yet one might wonder how this is possible in view of the discrepancy between the structures of extended substance and the mechanist world. A moving body may be composed of bodies individuated by their motions which are numerically distinct from it. But it is not possible that one extended substance has extended parts numerically different from it for reasons I have discussed. Mechanist bodies can be related as wholes and parts and extended substances cannot have substantial parts, yet the extensions of the former are extended substances. This can seem to be a problem. There may be a simple solution although it is not spelled out in the texts. They obscure both the problem and the solution as I see it.

The solution I have in mind is this. We might say that the extension of a whole is made of the extensions of its non-overlapping parts; that is, the whole is composed of the latter in continuous array. Consider a moving body which has no moving parts. Its constitution requires only the one extended substance whose boundary coincides with the path of the body. Now consider a body which has moving parts such that each of these parts either has no parts or can be decomposed into moving parts which have no parts. Suppose each of the partless parts is constituted by a motion and the limited extended substance which is coincident with the surface of separation. Finally, suppose the whole moving body consists of its motion and the extended substances which are constituents of each of its partless parts (and no other extended substances). The solution is, then, that the extension contained in the whole body is made up of the extended substances contained in its partless parts.[33] This may do to explain the metaphysical constitution of all finitely extended mechanist bodies. But Descartes maintains that when particles of the first element collide, they may be divided into parts of 'indefinite smallness';[34] there is a series of ever smaller particles in which 'we are unable to discover a limit'.[35] This poses no problem if the parts are created by a series of successive collisions, as the text implies. It would pose a problem if a particle were, at some time, composed of indefinitely many moving particles, but it is not clear to me that Descartes wants to say that this does, or can, occur.[36]

[33] This turns on the point that the bounded extension contained in the whole body need not be regarded as a (one) extended substance.

[34] P 3.52.

[35] Descartes explains that he uses 'indefinite' rather than 'infinite' because the latter term applies to God and it is absurd to think a finite mind can understand the infinite. By saying that something is 'indefinite', he means that although he cannot prove, or even conceive, that it is limited, he does not know that it is not because God may have reasons that he does not comprehend (AT 5, p. 51; CSMK, p. 320); also AT 7, p. 113; CSM 2, p. 81; AT 5, pp. 274–5; CSMK, p. 364.

[36] If he does, a solution to the problem in view will depend on clarification of the notion of 'indefinite smallness'.

Once the point that a mechanist body is individuated by a motion, not by a substance, is in view, the question of the relation between such a body and extended substance(s) looms. It is more complicated than the mode–substance relation seems to be on the determinate–determinable model. This is not apparent from the texts; they openly confront neither the condition-relative status of extended substances nor how some of them are selected as constituents of extant mechanist bodies. The difference between actual bodies and extended substances is papered over by Descartes's use of the key terms 'matter' and 'body'. In the passage that announces the one-motion, one-body rule, which I have quoted, motions are ascribed to 'parts of matter'; this is naturally understood to mean quantities of extension (substances), but the point of the passage is that motions belong to mechanist bodies which, as we know, are not just the same as extended substances.[37] Again: 'the term "body"... means a thing which has length and breadth and extension'.[38] Both a corporeal substance and a ship, for instance, satisfy this definition. By and large, the metaphysically important distinction between substantial and non-substantial bodies is ignored in Descartes's writings.

To take another example: '... there is no difference between the extension of a space, or internal place, and the extension of a body. For a body's being extended in length and depth in itself warrants the conclusion that it is a substance, since it is a complete contradiction that a particular extension should belong to nothing' (P 2.16; AT 8, p. 49; CSM 2, p. 230). This obscures the point that a mechanist body can, and an extended substance cannot, be composed of parts numerically different from it. To be sure, it is necessary that an extension belongs to (is) a corporeal substance rather than nothing, according to Descartes, but nothing in the passage, or in Cartesian metaphysics, warrants the conclusion that a body to which extension belongs is identical to a substance. The suggestion to the contrary omits complications perhaps for the sake of expository simplicity. The quotation is taken from an article in *Principles* 2 which precedes introduction of the complicating factors, 'motion' in the strict sense and the one-motion one-body rule.

Other opportunities to speak accurately about the constitution of bodies are neglected. Referring to the elements of the scholastics, the chemists' *tria prima*, salt, sulphur, and mercury, and easily convertible sorts of things such as water, ice, and snow, Descartes says: 'I think all these bodies are made of the same matter, and that the only thing which makes a difference between them is that the tiny parts of this matter which constitute some of them do not have the same shape or arrangement as the parts which constitute the others.'[39] Metaphysically speaking, the passage allows either of two interpretations: on one, a minimum part of mercury is a substance qualified by a certain figure and arrangement of its parts. On the other, it comprises modes and one or more substances subordinated to the modes; that is, although the modes reside in

[37] Also Descartes to Mersenne, 9 January 1639; AT 2, p. 482: CSM 1, p. 132.
[38] Descartes to Mersenne, 28 October 1640, AT 3, pp. 213–24: CSMK, p. 155.
[39] Descartes to the Marquess of Newcastle, 23 November 1646; AT 4, p. 570: CSMK, p. 302.

the substance(s), the substances are secondary in the constitution of particles or mercury. Because the substance(s) are made actual by uniting with the actual modes, the second interpretation is more suitable.

There is, however, one much discussed passage which highlights the distinction between naturally destructible bodies and extended substances:

> First, we need to recognize that absolutely all substances...are by their nature incorruptible and cannot cease to exist unless they are reduced to nothingness by God's denying his concurrence to them. Secondly, we need to recognize that body, taken in the general sense, is a substance, so that it too never perishes. But the human body, in so far as it differs from other bodies, is simply made up of a certain configuration of limbs and other accidents of this sort.... [A] human body loses its identity merely as a result of a change in the shape of some of its parts. And it follows that...the body can very easily perish...
>
> ('Synopsis' to *Meditations*, AT 7, p. 14: CSM 2, p. 10)[40]

In accord with the interpretation urged in this chapter, 'body in the general sense' is best taken to mean substantial bodies considered in general, or extended substance *as such*. Some scholars take it to mean the unique extended substance, ignoring or dismissing *Principles* 1.60. They typically take the passage to indicate that the one *res extensa* is to a human body as a substance is to a collection of modes. By implication, all composite bodies are nothing but constellations of modes.[41] This avoids the conundrum, but it cannot be exactly right. This is because all mechanist bodies are extended and impenetrable, but neither extension nor impenetrability is a mode. Colliding bodies behave like quantities of extension. This cannot be explained if bodies are nothing but collections of modes. Its explanation requires bodies to comprise numerically distinct extended substances because impenetrability is a multi-place relation and a consequence of the essential substantiality of extension. The behaviour of mechanist bodies is easily explained on the view that they are jointly constituted by motions and limited extended substances. In fact, the passage just quoted does not say that the human body is a collection of modes, but just that it differs from other bodies in respect of its modes.

To bring out how this account affects the ontological status of Cartesian bodies, we can compare it to the view of substances and modes propounded by Arnauld and Nicole in their treatise on logic. According to them, there are three sorts of entities that we conceive: a thing (substance), a manner of a thing (mode, attribute, quality), or a modified thing. With regard to the latter, 'I call something a "modified thing"

[40] Elsewhere, Descartes uses 'the body of a man' to refer to whatever body is in union with the soul of a man; as such, the body cannot be naturally destroyed because it gets its identity from that of the substantial soul; see to Mesland, 9 February 1645; AT 4, p. 167; CSMK, p. 243; also to Vorstius, 19 June 1643; AT 3, pp. 686–8; CSMK, pp. 224–5; Fourth Replies, AT 7, p. 250; CSM 2, p. 174.

[41] See Keeling, *Descartes*, 280–1; Dicker, *Descartes*, 295–301. Neither raises the question why composite bodies are extended and impenetrable in collision. They argue that Descartes's theory of the material world is untenable for the reason that motion apart from force is insufficient to differentiate parts of homogeneous extended matter; see note 25.

when I consider the substance as determined by a certain manner, or mode'; e.g. 'when joining the mode [roundness] with the substance, I consider a round body, this idea represents a modified thing'. Moreover, 'Names which signify things as modified, marking the substance primarily and directly although more confusedly, and indirectly marking the mode although more distinctly, are called adjectives... such as "round", "hard", "just", "prudent".'[42] On this view, a moving body is a substance and motion is added as a qualifier confusedly signified by an adjective. This comports with Descartes's general account of the substance–mode relation, but on the account of his ontology at which we have arrived, a mechanist body is not mainly constituted as a substance and only fully determined, or adjectively affected, by motion. Instead, a mechanist body is an entity framed by a motion and made of substance(s) as a statue is made of clay or a table, of wood.[43]

Putting metaphysical constitution aside, Descartes maintains that in mechanist theory, all bodies are composed of other bodies. Comparing his assumptions to those of competing explanatory theories, he boasts of

> ...my single assumption that all bodies are composed of parts. This is something which is visible to the naked eye in many cases and can be proved by countless reasons in others. All that I add to this is that the parts of certain kinds of bodies are of one shape rather than another. This in turn is easy to demonstrate to those who agree that bodies are composed of parts.
> (Descartes to Morin, 13 July 1638; AT 2, p. 200; CSMK, p. 107)

Parts can be defined in accord with the one-motion one-body rule, and I am aware of no other basis on which actual parts of matter are distinguished; figure is a candidate, but figure is due to motion.[44] If a body has no moving parts, it is not clear how parts of it are to be individuated. In practice, Descartes sidesteps this issue by positing three elements which are distinguished, more or less directly, by their motions:

> The first element is made up of matter which is so violently agitated that when it meets other bodies it is divided into particles of indefinite smallness.... The second is composed of matter divided into spherical particles which are still very minute when compared with those that we can see with our eyes, but which have a definite fixed quantity and can be divided into other much smaller particles. The third element... consists of particles which are much bulkier or have shapes less suited for motion. From these elements, as we shall show, all the bodies of the visible universe are composed. (P 3.52; AT 8A, p. 105: CSM 1, p. 258)

[42] *La Logique ou l'Art de Penser*, 5th edn, Part 1, ch. 2 (my translation).

[43] '[I]t is impossible to conceive of one part of an extended thing penetrating another equal part without thereby understanding that half the total extension is taken away or annihilated; but what is annihilated does not penetrate anything else; and so, in my opinion, it is established that impenetrability belongs to the essence of extension and not to that of anything else' (Descartes to More, 15 April 1649; AT 5, p. 342: CSMK, p. 372). As I understand it, the total extension of the two bodies would be diminished by the extension of the overlap; Descartes assigns half of this extension to one body and half to the other.

[44] See P 2.23; also Descartes to Villebressieu, Summer 1631; AT 1, pp. 216–17; CSMK, p. 33.

The first element pervades the heavens and the indeterminately small interstices between grosser bodies; the sun, stars, and other luminous bodies are composed of the second element; sensible terrestrial bodies are composed largely of the third. The three elements are inter-transumatable because larger particles shed tiny particles and small particles coalesce whenever one abuts another and the two are mutually at rest.[45] '[S]ome terrestrial particles continually take on the form of subtle matter when you crush them up; and some particles of this subtle matter attach themselves to terrestrial bodies so that there is no matter in the universe which could not take on all the forms one after the other.'[46] The phrase 'can take on all forms' echoes the scholastic formula that prime matter is the potential to receive all forms, but for Descartes 'matter' refers to corporeal substance and 'forms' refers to mechanist modes—motions, figures, and compositions of them. Like other mechanists, Descartes uses 'form' to refer to any combination of qualities that defines a sort of material things, e.g. water, fire, human body, and perhaps ships, clocks, and the like.

With the notion of forms of various material things in view, Descartes distinguishes between changes that a thing of a certain kind survives and those that destroy it: 'A simple alteration is a process which does not change the form of a subject, such as the heating of wood; whereas generation is a process which changes the form, such as setting fire to wood. Although both kinds of process come about in the same way, there is a great difference in the way of conceiving them and also in reality. For forms, at least the more perfect ones, are collections of a number of qualities with a power of mutual preservation. In wood there is only moderate heat to which it returns of its own accord after being heated; but in fire there is a strong heat, which it always preserves as long as it is fire' (Descartes to Regius, December 1641; AT 3, p. 461: CSMK, p. 201).[47] Animals and plants, which are preserved by a functional unity among their parts, are constantly gaining and losing particles, yet continue to exist apparently in virtue of the continuation of their forms. The distinction between these two sorts of changes is familiar and it has theoretical basis in scholastic-Aristotelian metaphysics, but it has no ground in Descartes's metaphysics and no clear ground in his physics. Descartes seems never to say that a division of material things into kinds has a real basis in the mechanism of the world. Some texts suggest the distinction of kinds is due to human attitudes and interests.[48]

In sum, for Descartes, modes are primary constituents of mechanist bodies, but they cannot be their sole constituents because something must account for their quantities of extension and mutual impenetrability, as we have seen. Explicating the ontological make-up of sensible things would involve intricacies which are not discussed in the text. What is perspicuous in Cartesian metaphysics is the theory of substances and modes which underpins the demonstration of mind–body dualism

[45] See P 2.55 on rest as the only glue.
[46] Descartes to Mersenne, 9 January 1639: AT 2, p. 485: CSMK, p. 133.
[47] Also see *The World*, ch. 5, AT 11, p. 26; CSM 1, p. 89.
[48] Normore, 'Descartes and the Metaphysics of Extension', pp. 285–6.

and can be modelled on highest determinables and their determinants. That makes it look as if substance is a single subject adjectively qualified by various combinations of modes, yet it is not possible that mechanist bodies are metaphysically constituted in this way, as I have argued in this section. Mechanist bodies are constituted in accord with mechanist physics, but no fully worked out theory of the basis on which material things are sorted into kinds, or the conditions in which they remain the same in change, is offered. Locke, by contrast, does his metaphysical work in this area. While he maintains a reasoned scepticism about the essence of corporeal substance and Descartes's real distinction of thinking and extended substances, he accepts a revisable mechanist theory as the most intelligible of the available explanatory hypotheses. His metaphysical interest lies where our knowledge does, as he sees it. That is in the sensible world of material things which are evidently composed of insensible particles, as Locke has it, whether or not they exactly conform to mechanist theory.

2. Locke on Modes, Bodies and 'Particular Kinds of Substances'

Rather like Descartes, Locke grounds our knowledge of substance on what we perceive and infer from this. We perceive recurring clusters of sensible qualities, according to Locke; we infer 'some *Substratum*, wherein they do subsist, and from which they do result, which therefore we call *Substance*' (E II. xxiii. 1). Substance is conceived in terms of what it does: it provides support for the existence of the qualities we find combined in one thing and it is unsupported itself.[49] We know in this extremely abstract way what its function is, but lack the ready conceptual knowledge of its nature claimed by Descartes. The idea of 'pure substance in general' is, Locke says, 'the supposition of [one] knows not what support' of Qualities 'commonly called Accidents' (E II. xxiii. 2). The idea of pure substance is an ingredient in all our ideas of substances, each of which also contains several ideas of sensible qualities which are supposed to belong to the same thing.

Yet it is not the case that the qualities included in all our ideas of substances are related to substratum in just the same way, according to Locke. The qualities represented in ideas of certain kinds of substances depend less directly on substratum than do the qualities represented in the most basic sorts of substances:

> If any one should be asked, what is the subject wherein Colour or Weight inheres, he would have nothing to say, but the solid extended parts: And if he were demanded, what is it, that that Solidity and Extension inhere in... [he could say only] 'something I know not what'.
>
> (E II. xxiii. 2)

Colour and weight inhere in solid extended parts; the solidity and extension of the parts inhere in substratum; colour and weight depend on substratum, too, but less

[49] Self-subsistence is ascribed to substances in E II. xii. 6.

directly than the extension and solidity of particles. Qualities of gross, sensible objects immediately inhere in a subject which is composed of several parts; the qualities of the parts inhere in substratum. Inherence is transitive, as Locke has it; so all qualities inhere in substratum. Some passages describe ideas of lead, a lodestone, and the like as collections containing ideas of certain qualities and the idea of substratum.[50] This omits the intermediate particulate subject of inherence but it is true that the idea of a lodestone or lead contains the idea of substratum.

The passage just quoted is significant for what it implies about the character of the inherence relation and why we know so little about substratum, according to Locke. The first answer is pertinent to the question: what do sensible qualities inhere in? Plainly enough, the answer invokes the mechanist hypothesis that the sensible qualities of bodies are explicable by the qualities we conceive to belong universally to all material things. Locke calls them 'primary qualities' and lists extension, figure, size, motion, position, and solidity as examples.[51] There is little reason to doubt that the second inference has a similar motivation, the felt need that the solidity and extension of the parts be explained by something. For by Locke's lights, the attributes taken as primitive in Cartesian mechanist theory are in need of explanation themselves. Unconvinced by Descartes's contention that extension, extended substance, and impenetrability are conceptually inseparable, Locke argues that our idea of spatial extension is different from our idea of body, which is a sort of substance. The idea of body includes not just extension but also solidity, both of which inhere in unknown substratum. As he sees it, bodies resist mutual penetration, fill space, and communicate motion because of their solidity, and we are far from understanding the force, or power, which has these effects (E II. iv. 6). Nor do we understand how solid extended parts cohere together especially in atoms whose parts are naturally inseparable (E II. xxiii. 23–7).[52] Substratum would, then, seem to be that entity (or complex of entities) which, if we knew enough about it, would give us a complete explanation of the solidity and extension of particles; according to mechanist theory, this would, in turn, be the basis for explaining colour, weight, and all other sensible qualities which belong to bodies composed of particles. Of course the mechanist hypothesis might not be entirely correct. But in any case, for Locke, the property of self-subsistence that pertains to a substance is explanatory adequacy. Substratum is supported by nothing because it is, or would be if we knew it, sufficient to explain all qualities and powers of material things.

With regard to minds, we conceive all minds to engage in thinking, perceiving, and willing. Like the universal attributes of matter, resistance, cohesion, and extension, the mental attributes are said to depend directly on substance: '... Thinking, Reasoning, Fearing, *etc.* which we concluding not to subsist of themselves, nor apprehending how

[50] E.g. 'Thus if to Substance be joined the simple *Idea* of a certain dull whitish colour, with certain degrees of Weight, Hardness, Ductility, and Fusibility, we have the *Idea* of Lead' (E II. xii. 6; also II. xxiii. 4 and 6).

[51] A few lists include number and texture.

[52] See Woolhouse, 'Locke and the Nature of Matter'.

they can belong to Body, or be produced by it, we are apt to think these the Actions of some other *Substance*, which we call *Spirit*' (E II. xxiii. 5). According to Locke, we have 'no clear and distinct idea of the *Substance* of Matter', nor any 'notion of the *Substance* of Spirit' (E II. xxiii. 5; also 15–31). Someone may say that 'he knows not what 'tis thinks in him' meaning 'he knows not what the substance is of that thinking thing'; again, he says that 'he knows not how he thinks'. Neither does he know 'what the substance is of that solid thing' or 'how he is extended; how the solid parts of Body are united, or cohere together to make Extension' (E II. xxiii. 23). Notice the direct move from primary ideas of spirits and bodies to substratum; for Locke, what we don't know about the causal origin of the cohesion of parts and resistance to penetration, and what we don't how about the causes of acts of thinking and willing, is just what we don't know about substratum. Descartes claims to know what is sufficient for thinking, on the one hand, and impenetrable extension, on the other. Locke denies that we have this knowledge. If we did, we would know what their substrata are.

In addition to ideas of body and spirit, we have ideas of 'distinct sorts of Substances', as Locke calls them. These are collections of ideas of sensible qualities which are observed 'to exist together, and are therefore supposed to flow from the particular internal Constitution, or unknown Essence of that Substance' (E II. xxiii. 3). By contrast, the attributes represented by our primary ideas of body, that is, resistance to bodily penetration, cohesion of parts, and extension, and those represented in or basic ideas of spirits, that is, powers of thinking and moving the body at will, are never, to my knowledge, said to arise from, or inhere in, an inner constitution, or real essence of a kind. I take this to be because inner constitutions, as Locke understands them, presuppose insensible particles whose mutual resistance, cohesiveness, and extension subsist directly in substratum (E II. xxiii. 2). Examples of particular sorts of Substances include man, horse, sun, lead, opium, and the like; these are kinds defined by several sensible qualities that are combined in the same thing and result from the 'Bulk, Texture, and Figure of the minute parts'; that is, from their inner constitutions (E II. xxiii. 8). Consider gold, for instance: '…it is the real Constitution of its insensible Parts, on which depend all those Properties of Colour, Weight, Fusibility, Fixedness, etc. which are found in [a parcel of gold]' (E III. iii. 18; also II. xxiii. 6, 8, II. xii. 6).

Ontologically speaking, gold consists of a number of insensible particles, each of which is a substance, and several entities which are not in the category substance, the primary qualities of the several particles and a shared texture. Yet Locke says that antimony and lead, for example, are substances of different kinds. It is worth asking why they deserve to be called 'distinct particular sorts of Substances' (E II. xxiii. 3). After all, they are not different with respect to their substance; on the contrary, their respective substances are of exactly the same kind, extended solid particles of matter. The case is different with body and spirit. The solidity and extension by which we define body inhere directly in substratum; and thinking and willing inhere directly in a substratum which may, for all we know, be of a different kind—immaterial rather than material. But gold, water, horse, and the like are the same in respect of their substance and differ only with regard to the primary qualities and textures of their substantial parts. So one might wonder: how does Locke's tendency to say that a horse and a violet, for example, are distinct kinds of substances differ from saying that a brick wall and a brick oven are distinct kinds of bricks?

It seems plain that Locke speaks as he does because of the evident substance-like properties of horses, violets, gold, and the like. As we saw, the similarity between sensible qualities like colour and weight which result from a multitude of insensible particles, on one hand, and the extension and solidity of the particles which depend on substratum, on the other, is introduced at the start of the well-known chapter 'Of the Complex Ideas of Substances'. Both inner constitutions and substratum support combinations of qualities, but we are forewarned against supposing they are 'substances' in just the same sense. Substratum is called 'substance' in the first instance because, as the text explains, we suppose qualities subsist in it (E II. xxiii. 1); all qualities subsist in it, indirectly if not directly. It seems, then, that gold and a horse are called 'substances' because they, too, support combinations of qualities although in a derivative way. Locke's contemporaries could easily have recognized this as 'substance' in an analogous sense.[53] So we might say that the substratum is substance to the solidity and extension of particles and the inner constitutions of gold, horse, and so on are substance analogues.

Still many scholars take the expression 'different kinds of substances' to refer to kinds of corporeal substances, as if 'corporeal substance' signified a genus. Aristotelian substances are ordered by genera and species; corporeal substance is a genus under which there are several species. But Locke favours a version of the mechanist theory on which there is only one kind of material substance; it is varied by a number of qualities and modes that subsist, directly or indirectly, in substrata of the same kind.[54] Nevertheless, the relation between corporeal substances and kinds of material things such as mercury, gold, and horses may be obscured by the determinable–determinant picture of substance and mode in familiar Cartesian texts. This is too simple as an account of the constitutions of Cartesian mechanist bodies, but this is not brought out by Descartes. Readers of Locke may, then, be primed to overlook the crucial distinction between particulate constitutions and substratum outlined in *Essay* II. xxiii. 2, the difference between substances and their analogues. Accordingly, they suppose Locke takes gold, horse, and the like to be substances on a par with solid extended parts. To be sure, substratum is an element in all sorts of material things, an element unfailingly represented in the ideas of body, spirit, lodestone, horse, and the rest. But the diversity of kinds is at the level of material things with inner constitutions, not the

[53] Although 'substance' has a sort of ambiguity in Locke's usage, it is of the sort scholastics ascribed to 'analogous terms', words which have a primary signification and a derivative one in which they are applied to things which are related to what they primarily stand for by causality, similitude, or being a sign of. For the doctrine in the seventeenth century, see Arnauld and Nicole, *La Logique ou l'art de penser*, I, 6.

[54] Bolton, 'Locke's Account of Substance in Light of his General Theory of Identity', defends this interpretation of E II. xxiii, related chapters of E III, and E II. xvii 'Of Identity and Diversity' in more detail.

level of insensible particles. Overlooking this articulation of Locke's theory of substance wrecks havoc with his metaphysical theory of composite entities, as we will see.[55] It allows the *substance* comprised in the constitution of, say, a horse to be regarded as a subject adjectively determined by the qualities typical of horses; the *inner constitution* is then reduced to an aggregate of solid coherent particles and identified with substratum. Structural elements, textures, and arrangements which unify stable collections of particles and ground their joint causal powers are left out of account. In fact, Locke's ontology is not this spare.

Locke designates the basic entities in his ontology by reviewing the sorts of ideas we have. These are ideas of substances, modes, and relations.[56] There is not space here for a proper account of relations or modes, as Locke understands them; an overview will have to do. Relations, according to Locke, are ways of comparing, or considering, two things while keeping them distinct. Relations depend upon their *relata*; they are founded in intrinsic features of the things they span, and if any one of them is altered or destroyed, the relation ceases to exist (E II. xxv. 5). Although relations depend on substances, they do not typically do so in just the same way as the qualities which belong to substances do. As Locke puts it, 'This farther may be considered concerning *Relation*, That though it be not contained in the real existence of Things, but something extraneous and super-induced: yet the *Ideas* which relative Words stand for, are often clearer... than those Substances to which they belong' (E II. xxv. 8). 'Things' refers to substances. Paternity, for example, is extraneous to a man, because it is not included in our idea of man, which of course does include the various qualities we deem distinctive of human beings.[57]

Yet the idea of extension, which is classified as a simple idea, contains relations among ideas that represent relations among extended parts. With reference to space

[55] An anonymous reader asked whether this isn't encouraged by passages that say the qualities of, say, gold inhere in and depend upon the primary qualities (bulk, figure, size) of its solid parts (e.g. II. viii. 16; II. xxxi. 6). I do not think this is intended by Locke. It is true that such passages stress limits on our conception of what qualities of gold depend on—we can only conceive that they depend on primary qualities of its parts. This does not mean the primary qualities depend on nothing more basic, but that we have no conception at all of what that might be (E II. xxiii. 22–3). Moreover, such passages call attention to subjects at two levels—sensible things with a variety of qualities and insensible parts with primary qualities. The reasoning that leads from the existence of qualities to the existence of something in which they subsist applies at both levels and arrives at different subjects of inherence at each. Descartes would say otherwise because he takes extension to be the essence of corporeal substance. Extension is ontologically prior (has more reality) and all things in the category modes are posterior ('Second Set of Replies', AT 7 p. 165: CSM 2, p. 117).

[56] E II. xxvii. 2. Locke's use of 'mode' and 'relation' is irregular. In some contexts, relations are explicitly subsumed under modes for purposes of discussion (E II. xxx. 4; II. xxxi. 14). 'Mode' is the word early modern philosophers typically use to refer to all things which depend for their existence on substances. Locke uses it that way in his letters to Stillingfleet, but the *Essay* begs pardon for giving it a signification different from the usual one (E II. xii. 4).

[57] Ayers, *Locke II: Ontology*, 94–5. Some scholars take this passage and a few others to show that Locke maintains that relations are not real, but mind-dependent acts of referring or comparing. Although this view deserves discussion, I disagree. One reason is that relations are on Locke's list of 'things there are' (E II. xxvii. 2). I am grateful to Jean-Michel Vienne for allowing me to read his thoughtful unpublished paper on this topic.

and duration, 'none of the distinct *Ideas* we have of either is without all manner of composition, it is the very nature of both of them to consist of Parts' (E II. xv. 9).[58] Solidity is also relational, a power of one body with respect to other bodies.[59] Sensible qualities in general turn out to comprise relations:

> For our Ideas of Extension, Duration, and Number, do they not all contain in them a secret relation of the Parts? Figure and Motion have something relative in them much more visibly: And sensible qualities, as colours and Smells, etc. what are they but the Powers of different Bodies, in relation to our Perception, etc. And…do they not depend on the Bulk, Figure, Texture, and Motion of the Parts? All of which include some kind of relation in them.
>
> (E II. xxi. 3)

Locke gives the name 'ideas of relation' to compound ideas that contain ideas of two things and a way they are related, but it seems that simple ideas of qualities can be unpacked as ideas of relations if need arises.

Modes are the final item in Locke's ontology, and they, too, require more discussion than they get here. In brief, they, too, depend for their existence on substances. There are simple modes (e.g. numbers, a triangle) and mixed modes (e.g. beauty, theft, a dance).[60] Many, if not all, modes are in effect determinates of a determinable dimension. Ideas of modes are compounded of a base simple idea which is modified by combination with one or more other ideas. It seems that the base idea represents a determinate. This supposedly makes it suitable to be combined with certain other ideas in ways that result in ideas of other determinates of the same determinable…[61] Ideas of simple modes are formed from ideas of the same kind (same determinable dimension); they include numbers, geometrical figures, measures of spatial and temporal distance, motions, shades of colour, tones, and other secondary qualities. Ideas of mixed modes are formed from ideas of more than one kind. Many mixed modes are sorts of actions; for this reason, Locke explains, the simple ideas which have been most modified and included in ideas of mixed modes are ideas of motion, thinking, and power.[62] In general, Locke maintains that ideas of modes are 'made arbitrarily', that is, they are not, as such, implicitly referred to patterns that exist in nature.[63] Yet some of them are derived by observing performances such as fencing or wrestling; others are invented in the absence of things of the sort they specify, as Locke supposes the idea of adultery may have been.[64] But all ideas of modes are intended to represent sorts of things that could possibly exist, and are defective if they do not.[65]

[58] See note to this section added in the fifth edition. [59] E II. iv. 1–3. [60] E II. xii. 4–5.

[61] E II. xiii. 15–18. On simple modes of the idea of space, see Carson, 'Locke on Simple and Mixed Modes', pp. 19–38.

[62] E II. xxii. 10. [63] E II. xxx. 4; II. xxxi. 3; these texts say the same thing about ideas of relations.

[64] E II. xxii. 9; E III. v. 3.

[65] In general, an idea is said to be 'real' just in case it conforms to its archetype, i.e. the standard to which it is 'tacitly referr'd'; applying this to the idea of a mode, it is said to be real provided that it is possible for something that conforms to it to exist (E II. xxx. 1 and 4).

The ontological status of modes is relevant to our topic only insofar as modes are elements in the constitutions of sensible material things. Locke's description of the components of living things indicates that vital processes are classed as modes. Yet natural processes such as fermentation, chemical reduction, digestion, and procreation do not fit comfortably in the category of either simple or complex modes. It hardly seems that sorts of things that are not constituted by nature are proper subjects of biological and medical enquiry. These issues cannot be pursued in this chapter,[66] but there is good textual evidence that Locke wants to say that modes are among the entities in the make-up of plants and animals.

In the *Essay*'s second edition, Locke added a chapter 'Of Identity and Diversity', apparently at a friend's suggestion that he write something on the *principium individuationis*.[67] Other chapters of Book II purport to describe a system of ideas shared by all human beings, but this one exposits the ontology Locke favours in an effort to resolve difficulties about the individuation of a thing at different times (E II. xxvii. 1). At the outset, it states that 'We have *Ideas but* of three sorts of Substances; 1. God. 2. Finite Intelligences. 3. *Bodies*' (E II. xxvii. 2; emphasis added).

Substances are said to be individuated by the kinds of things they are and the times and places at which they exist: '[T]hough these three sorts of Substances, as we term them, do not exclude one another out of the same place; yet we cannot conceive but that they must necessarily each of them exclude any of the same kind out of the same place' (E II. xxvii. 2). For instance, two bodies cannot exist in the same place at the same time. The passage goes on to say: 'All other things being but Modes or Relations ultimately terminated in Substances, the Identity and Diversity of each particular Existence of them too will be by the same way determined' (E II. xxvii. 2). An exception is made for things whose 'Existence is successive', such as motions and thoughts, for the reason that a succession contains nothing that exists at different times. The passage just quoted announces Locke's favoured ontology.

It also states that modes and relations are individuated on the same basis as substances: by their kinds and the times and places at which they exist. It is significant that they are not individuated by direct reference to the substance(s) on which they depend. We can presume that a mode or relation and the substance(s) to which it belongs are co-located in space and time. The point evidently is that different substances can occupy those positions at different times. The principle of individuation allows that one and the same relation (mode) might depend on a succession of different substances. This may not be possible for all of them, but it may be for some. For instance, if a dance goes around in a circle and fresh dancers replace tired ones over a period of time, the same numerical circle and dance exist at every moment during this period of time. The account is generalized to all sorts of things: 'the *principium Individuationis*...is Existence itself, which determines a Being of any sort to a particular time and place incommunicable to two Beings of the same kind' (E II. xxvii. 3).

[66] Anstey, *John Locke and Natural Philosophy*, 118–35 is helpful.

[67] 'Mr. Molyneux to Mr. Locke', 2 March 1693, LW 9, p. 310.

To understand what this implies about the individuation of a thing at different times, according to Locke, we need to consider the kind of thing it is. In the category substance, a Body is said to be any particle of matter 'to which no Addition or Subtraction of Matter [is] made'. Atoms, which have parts which are never separated in nature, are Bodies; so are masses of atoms, that is, two or more atoms 'united together', 'be they never so differently jumbled'.[68] A mass of atoms is a collection of atoms that cohere; as such, it comprises no particular relation (e.g. texture) or mode. Sensible things such as gold and mercury, which have inner constitutions but are stuffs rather than countable individuals, are not mentioned in this chapter. But it does undertake to explain the individuation of plants and animals.

A brief review of his account of the trans-temporal identity of several kinds of things is in order. An atom continues to exist as long as it has a position along a spatially and temporally continuous path and there is an atom at every point on this path. A mass of atoms continues to exist just as long as none of its atoms is gained or lost. Turning to living things, Locke observes that they remain the same even though they gain and lose great masses of matter; accordingly, their identity does not depend on 'a mass of the same particles'.[69] Suppose the kind of thing in question is an oak tree. To discover what its identity depends on, Locke specifies what an oak is in addition to a coherent mass of particles. An oak is 'such a disposition of [coherent particles] as constitutes the parts of an Oak; and such an Organization of these parts, as is fit to receive, and distribute nourishment, so as to continue, and frame the Wood, Bark, and Leaves, *etc.* of an Oak, in which consists the vegetable Life' (E II. xxvii. 4). The tree is an organization, a relation, among functional parts; each of the parts is constituted by a disposition, or arrangement, of particles; the vegetable life would seem to be an activity which depends on this complex of substances, arrangements, and organizations. In the phrase used by Locke in summarizing his overall theory of identity, an oak is 'the composition of distinct Substances, and different Modes' (E II. xxvii. 28).

Locke's theory is that an oak remains numerically the same as long as the life of an oak exists constantly 'in the same continuity of insensibly succeeding Parts united to the living Body of the Plant' (II. xxvii. 4). The trans-temporal identity of the tree consists in the continuation of a mode and a relation—a vegetable life and a functional organization. In effect, an oak is built around the organization which recruits gradually changing collections of particles which successively compose its operational parts. The life of the oak is carried on moment by moment by the sequence of different masses of particles which are party to the organization. An oak is not reducible to a collection of substances qualified by primary qualities. One reason is that the substances change but the oak remains numerically the same. More than this, certain structural modifications are main elements in the oak's metaphysical make-up. By 'structural modification', I mean a mode or relation that depends on two or more substances at a time and can depend on different substances at different times. An oak is constituted as an entity

[68] E II. xxvii. 3. [69] E II. xxvii. 3.

by structural modifications which unite with series of masses of particles suited to terminate the modifications for some period of time.

An oak is not a substance, but only an analogue of one. Many scholars assume, wrongly as I think, that Locke is committed to regarding it as a body, that is, a corporeal substance. But some take the following remark to be evidence of a distinction between two sorts of corporeal substances: 'though [the *principium individuationis*] seems easier to conceive in simple Substances or Modes; yet when reflected on, is not more difficult in compounded ones' (E II. xxvii. 3).[70] This is taken to mean that a distinction between simple and compounded things divides Locke's categories substance and mode. On one such interpretation, atoms are ranked as simple substances, whereas masses of particles and living things are classed as compounded substances. Yet Locke accounts for the metaphysical properties of an oak in a way that is entirely unsuited to a substance. A substance is individuated as a substance, and its attributes, qualities, and modes are unified by a substance. In Locke's list of items in the make-up of an oak, there is nothing that plays this role. The tree contains many and changing substances, remains the same over time in virtue of a mode, and is unified at a time by a relation. The remark just quoted, about the individuation of compounded substances and modes, is best understood either as an allusion to substances and substance analogues or to the composite entities mentioned in the conclusion of the chapter on identity (E II. xxvii. 28).

If an oak were a substance, Locke's description of its components would be inconsistent with other doctrines he holds. For *Essay* II. xxvii recognizes only three sorts of substances and an oak would have to be classified as a body; the oak and a mass of particles are co-located for some period of time; they are numerically distinct because the oak survives gain and loss of particles. The conjunction of these three conditions is inconsistent with the basic principle that two bodies cannot exist in the same place at the same time.[71]

Some scholars who understand this very well say that the plant is not a substance, but a mode.[72] It is true that it is individuated over time by a mode, but it cannot be strictly identical to a mode. This is because a tree incorporates successive masses of particles. Unlike a mode, an oak is extended and solid. Much the same might be said of a Cartesian body although Descartes says little about the fact that a body comprises something other than a mode. Locke's explicit account of the items comprised in an oak avoids the problem of co-located bodies and the problem of non-material modes. An oak is 'the composition ... of distinct Substances, and different Modes'.[73]

[70] Chappell, 'Locke on the Ontology of Living Things and Persons'.

[71] Shoemaker, *Self-Knowledge and Self-Identity*, pp. 45–6, seems to have been the first to charge Locke with this inconsistency. Alston and Bennett, 'Locke on People and Substances' proposes that Locke avoids the inconsistency in virtue of a harmless ambiguity in his use of 'substance' which is close to the truth, as I see it. Other scholars say Locke avoids the problem by a fallacious ambiguity.

[72] Uzgalis, 'Relative Identity and Locke's Principle of Individuation', pp. 62–3; also 93, which nevertheless says that living things are necessarily composed of material particles.

[73] E II. xxvii. 28.

It comprises one and only one material substance at a time (a mass of particles) and it is individually distinct from any of these masses because its constitution contains structural modifications as well.

The thesis of this chapter is that Descartes and Locke share two anti-Aristotelian doctrines regarding the metaphysical status of substances, modes that depend on substances, and composite entities such as Bucephalus. Let me conclude by pointing to a few notable differences between their view. If we assume that matter is what all material things are composed of and what underlies changes in which such things begin and cease to exist, substance is matter according to both theories. In his published works, Descartes seems to do little to explain how sensible things are made of *res extensa* or to analyse how it serves as substratum for change. In connection with these functions, Descartes refers to extended substance as 'matter' of which salt, mercury, etc. are made, or 'matter' which can 'take on all forms'. This hardly differentiates *res extensa* from a common scholastic notion of prime matter.[74] Yet if the interpretation urged in this chapter is correct, Descartes maintains that *res extensa* has a structure with regard to parts that provides an elegant account of how it functions as matter. In brief, *res extensa* underlies change in which bodies begin and cease to exist because any portion of it can be conceived without parts and also conceived as having whatever distinct parts are needed for the formation and dissolution of all mechanist bodies and all sorts of material things. For his part, Locke is open and explicit about what material things are made of and how they arise and perish. He maintains that there are atoms of matter—naturally indivisible, immutable particles which compose all other material things and underlie the various changes to which they are subject. This may be adequate for all physical changes that can occur in an atomistic world, but it is not adequate to the mathematical physics envisaged by Descartes.

Locke's interest lies in explaining the metaphysical structure of the world as we experience it. He maintains that there are material substances, but our way of conceiving them is more important to us than they are in themselves. Locke's working metaphysics starts with atoms, corresponding roughly to the moving bodies with which Descartes's metaphysics stops. What is most salient in our experience, as Locke sees it, are sorts of sensible things that exhibit the unity distinctive of Aristotelian substances. The lengthy acccount of our ideas of particular sorts of substances is a central part of his effort to explicate this unity, in terms of inner constitutions and, then, in terms of substratum. The theory of identity developed in *Essay* II. xxvii is also designed to account for this unity. There is nothing corresponding to this in the Cartesian texts. They purport to explain the unity among qualities of mercury, salt, etc., and the processes carried on by plants and animals on the basis of physical theory, not metaphysics. Although Descartes speaks of different sorts of material things and the changes they

[74] See Descartes to Marquess of Newcastle, 23 November 1946, AT 4, p. 570: CSMK, p. 302 and Descartes to Mersenne, 9 January 1639; AT 2, p. 485: CSMK, p. 133 (see above notes 39 and 46). *The World*, AT 11, p. 33: CSM 1, p. 91, complains that prime matter is 'so thoroughly stripped of all its forms and qualities that nothing remains in it that can be clearly understood'.

can undergo, he offers no systematic account of their unity-making features. The one exception is the human body, which is substance-like because it is in union with a mind. For Descartes, material things may have a sort of functional unity, but they are composed of parts. A unity of composition is different in kind from the unity of nature distinctive of substances as Descartes sees it. He finds little analogy between them.[75]

References

Alston, William and Bennett, Jonathan. 'Locke on People and Substances', *Philosophical Review*, 97 (1988), 25–46.

Anstey, Peter. *John Locke and Natural Philosophy*. Oxford: Oxford University Press, 2011.

Arnauld, Antoine and Nicole, Pierre. *La logique ou l'art de penser*. Cinquième édition revue & de nouveau augmentée. Paris chez Guillaume Desprez, etc., 1683.

Ayers, Michael. *Locke II. Ontology*. London: Routledge, 1990.

Bolton, Martha. 'Thinking: The Nature of Descartes' Mental Substance', in Karen Detlefsen, ed., *Descartes' Meditations: A Critical Guide*. Cambridge: Cambridge University Press, 2013: 65–81.

Bolton, Martha. 'Locke's Account of Substance in Light of his General Theory of Identity', in Paul Lodge and Tom Stoneham, eds., *Locke and Leibniz on Substance*. New York: Routledge, 2015: 63–88.

Carson, Emily. 'Locke on Simple and Mixed Modes', *Locke Studies*, 5 (2005), 19–38.

Chappell, Vere. 'Locke on the Ontology of Living Things and Persons', *Philosophical Studies*, 60 (1990), 19–32.

Chappell, Vere. 'Descartes's Ontology', *Topoi*, 16 (1997), 111–27.

Dicker, Georges. *Descartes: An Analytical and Historical Introduction*, 2nd edn. Oxford: Oxford University Press, 2013.

Garber, Daniel. *Descartes' Metaphysical Physics*. Chicago: University of Chicago Press, 1992.

Grosholz, Emily. 'Descartes and the Individuation of Physical Objects', in Kenneth Barber and Jorge Gracia, eds., *Identity and Individuation in the Early Modern Period*. Albany, NY: State University of New York Press, 1994: 41–58.

Keeling, S. V. *Descartes*, 2nd edn. Oxford: Oxford University Press, 1968.

Lennon, Thomas. 'Descartes' Idealism', *Philosophie et culture*, 4 (1988), 53–6.

Lennon, Thomas. 'The Problem of Individuation among the Cartesians', in Kenneth Barber and Jorge Gracia, eds., *Identity and Individuation in the Early Modern Period*. Albany, NY: State University of New York Press, 1994: 13–40.

Lewis, Genevieve. *L'Individualité selon Descartes*. Paris: Vrin, 1950.

Lowe, E. J. *Locke*. London: Routledge, 2005.

Normore, C. G. 'Descartes and the Metaphysics of Extension', in Janet Broughton and John Carriero, eds., *A Companion to Descartes*. New York: Wiley-Blackwell, 2011.

Rozemond, Marlene. 'Descartes' Dualism', in Janet Broughton and John Carriero, eds., *A Companion to Descartes*. New York: Wiley-Blackwell, 2011.

[75] I am grateful to the participants and organizers of the Conference on Locke and Cartesianism held at the University of Lille in 2012, especially Michael Ayers and Philippe Hamou, as well as two anonymous readers, one of whom was particularly insightful.

Secada, Jorge. *Cartesian Metaphysics*. Cambridge: Cambridge University Press, 2000.

Shoemaker, Sidney. *Self-Knowledge and Self-Identity*. Ithaca, NY: Cornell University Press, 1963.

Stuart, Matthew. 'Descartes' Extended Substances', in Rocco Gennaro and Charles Huenemann, eds., *New Essays on the Rationalists*. Oxford: Oxford University Press, 1999.

Uzgalis, William. 'Relative Identity and Locke's Principle of Individuation', *History of Philosophy Quarterly*, 7 (1990), 283–97.

Williams, Bernard. *Descartes: The Project of Pure Reason*. New York: Penguin Books, 1978.

Woolhouse, Roger S. 'Locke and the Nature of Matter', in Eileen O'Neill and Christia Mercer, eds., *Early Modern Philosophy: Mind Matter and Metaphysics*. Oxford: Oxford University Press, 2005: 142–60.

6

Virtual Existence of Ideas and Real Existence

Locke's Anti-Cartesian Ontology

Matthieu Haumesser

Throughout the *Essay*, the idea of 'existence' scarcely comes to the forefront. This may be understood as a consequence of Locke's apparent rejection of metaphysics and ontology. His 'historical method', based on the empirical observation of the mind and its operations, doesn't leave much room for metaphysical suppositions, especially those related to substantiality. In the background of this way of thinking lies his criticism of Descartes, whose philosophy relied much more on material and spiritual substances, and accordingly on a full-fledged ontology.

In this chapter, however, I would like to show that the rare remarks about existence that are nevertheless made in the *Essay* are all the more interesting, even though they stay in the margins of Locke's discourse. They may be considered the lineaments of an implicit ontology, which would be quite significant as such. One may also assume that these remarks about existence are in Locke's mind certainly related to his interpretation of Descartes's philosophy. There is one—rather short—chapter in the *Essay* where Locke gives an explicit account of his conception of existence. It is chapter II. vii. 7, dealing with ideas that come from both sources of experience, that is, sensation and reflection. Locke considers here a whole set of ideas that originate in the consideration of the operations of the mind (reflection) as well as in the experience of external bodies (sensation): pleasure, pain, unity, duration, and existence. These ideas, and the idea of existence in particular, can therefore be considered both ideas of reflection and ideas of sensation. This dual nature, however, raises a serious question: does a body *exist* in the same way as an operation of the mind—or the mind itself?

My purpose here will be to demonstrate that Locke strongly distinguishes between two kinds of existence, even if they have something to do with each other. Locke constantly oscillates between the virtual, transient existence of ideas in the mind on the one hand, and the 'real existence' (of bodies, mainly) on the other hand. This double approach of existence determines Locke's rejection of Descartes's ontology and the elaboration of what one may call the ontological background of the *Essay*, based on this double meaning of 'existence'. I will eventually focus on Locke's use of metaphors, which allows him to establish a relationship between the intellectual world and the material world, as well as the two different kinds of existence that constitute them.

1. Existence of Ideas, Existence of Substances

Chapter II. vii gives a first overview of the various simple ideas that come from both sensation and reflection (i.e. from bodies, and from the mind considering its own operations). After the simple ideas of sensation like colours, solidity, extension, and the simple ideas of reflection, Locke comes to examine a number of ideas that originate in both sources of human knowledge:

There be other simple *Ideas*, which convey themselves into the mind, by all the ways of Sensation and Reflection, *viz.*
Pleasure, or *Delight*, and its opposite.
Pain, or *Uneasiness.*
Power.
Existence.
Unity.[1]

In this chapter, Locke doesn't really insist on the special status thus attributed to these ideas. It is clear however that they must have a transverse position in his analysis of human knowledge, as they apply at the same time to the external world of bodies and to the internal world of the operations of the mind. Their status is all the more puzzling as this transverse position—and more particularly the fact that they can be obtained by reflection—may be considered a good argument in favour of innate ideas. It is for example an essential feature of Leibniz's position: 'Ideas of reflection originate in our mind; I would like to know, how we could get our idea of being, if we weren't beings ourselves, and if we didn't find being in ourselves.'[2] According to Leibniz, the idea of existence (but the same could be said of power, unity, etc.) does not depend on any sensory experience, but is already implied in the very nature of our own understanding, and is therefore innate.

The same kind of argument was already developed in Descartes's *Meditations* and *Principles*, where a special status is given to ideas, such as 'existence' or 'substance', that can be attributed to both kinds of substances. Descartes makes this point in the Third Meditation in order to show that all the ideas constitutive of our knowledge of bodies may have been found originally in ourselves, in the perception of our own nature, and only afterwards 'transferred' to external things (or what common sense believes to be external things)—which would also mean that they are originally innate:

[1] E II. vii.1. [2] *Nouveaux Essais sur l'entendement humain*, A, 6. 6, I.1, §23, pp. 85–6.

With regard to the clear and distinct elements in my ideas of corporeal things, it appears that I could have borrowed some of these from my idea of myself, namely substance, duration, number and anything else of this kind. For example, I think that a stone is a substance, or is a thing capable of existing independently, and I also think that I am a substance. Admittedly I conceive of myself as a thing that thinks and is not extended, whereas I conceive of the stone as a thing that is extended and does not think, so that the two conceptions differ enormously; but they seem to agree with respect to the classification 'substance' ['in so far as they represent substances' (according to the French version)] (AT 7, p. 44: CSM 2, p. 30)

This agreement found between the two kinds of existence, and the possibility of 'transferring' the spiritual idea of existence (but also of number, and duration) to the world of bodies, are therefore strong points in favour of innatism. It should be noted that Descartes understands here existence—of spirits, and of bodies—as the existence of *substances*. That is a strategic feature of an ontology based on one unified concept of substance that must be common to spirits and bodies, even though the difference between the principal attributes of these beings (extension and thought) necessarily leads to a dualistic view of reality. Descartes is very clear about that in the title of §52 of the *Principles* (part 1): '*The term "substance" applies univocally to mind and to body.*'

Now, if we consider from this Cartesian point of view Locke's remarks on existence in II. vii, we find significant discrepancies. Like Descartes, Locke says that the idea of existence comes from the mind reflecting on its own thoughts, as well as from the bodies; but unlike Descartes, he does not infer from this observation that the spiritual idea of existence has more weight or evidence than the material one. He rather suggests the contrary: the existence perceived by reflection is only the existence *of Ideas*. There is no reference here to a thinking *substance*, which is of course a strong difference from the Third Meditation. It is all the more significant as sensation, on the contrary, provides the mind with the idea of the existence of 'things', or 'real beings':

Existence and *Unity*, are two other *Ideas*, that are suggested to the Understanding, by every object without, and every *Idea* within. When *Ideas* are in our Minds, we consider them as being actually there, as well as we consider things to be actually without us; which is, that they exist, or have *Existence*: And whatever we can consider as one thing, whether a real Being, or *Idea*, suggests to the Understanding, the *Idea* of *Unity*. (E II. vii. 7)

'Whether a real being, or *idea*': it is clear here that even if ideas do have existence, they do not exist as 'real beings'. Similarly, at the beginning of Book IV, Locke will also distinguish between the knowledge of our ideas (with regard to their identity, diversity, or relations), and the knowledge of 'real existence'. If we take these statements seriously, we shall conclude that according to Locke, the existence of ideas, such as perceived by reflection, cannot exactly be 'real existence'. Therefore, if not real, what kind of existence is it? And how are we to consider its relationship with 'real beings', if it is still true that the understanding deals with existence in both cases, and that both cases partake, after all, of one reality?

2. 'To be in the understanding' is 'to be perceived'

The rejection of innate ideas and principles, in the first book of the *Essay*, raises from the start the question of the kind of reality that shall be attributed to ideas. The problem of innatism, according to Locke, is precisely that it gives too much 'substance' to certain ideas, by considering them 'innate characters', 'stamped into the Mind' (I. ii. 1), that would therefore 'exist' in it, even when the mind does not actually perceive them. Such an existence would be independent of the mind using its intellectual powers, and this is precisely what Locke does not accept in this supposition. Therefore, he intends to show in the *Essay* (and especially in Book II) 'how Men, barely by the Use of their natural Faculties, may attain to all the Knowledge they have, without the help of any innate Impressions' (I. ii.1). According to Locke, the mind and its ideas exist not as substances or 'real beings', but always through a number of *operations* in which human faculties are set to work. It is clear that such an existence, in Locke's view, is much more valuable than the passive existence of substances. That is why the first operation of the mind is *perception*, where there is necessarily a minimal degree of active consciousness:

> For if these Words (*to be in the Understanding*) have any Propriety, they signify to be understood. So that, to be in the Understanding, and, not to be understood; to be in the Mind, and never to be perceived, is all one, as to say, any thing is, and is not, in the Mind or Understanding. (E I. ii. 5)

An idea therefore has no existence at all when it is not actually perceived. The intellectual world of the understanding that Locke describes in Book II entirely relies on this operation of perception. That makes an essential difference with the 'real existence' attributed to substances, and to bodies in particular, which consists precisely in the supposition that they continue to exist when we do not perceive them.

Unlike Descartes, Locke maintains throughout the *Essay* this strong difference between existence of ideas in the understanding and existence of material or spiritual substances. Let us consider the definition of simple ideas, at the beginning of Book II. In this passage, Locke clearly distinguishes how they exist in the mind from how qualities exist in substances:

> Though the Qualities that affect our Senses, are, in the things themselves, so united and blended, that there is no separation, no distance between them; yet 'tis plain, the *Ideas* they produce in the Mind, enter by the Senses simple and unmixed. (II. ii. 1)

The simplicity of these ideas is the most central feature of the intellectual world of the understanding. It is a direct consequence of the importance given to perception: the ideas that exist first and foremost in the mind are those that do not divide attention—therefore the simple ones. That is why they must be considered the 'material of all our knowledge'.

Paradoxically, this could also mean that these simple ideas, even if they are the very first to come into the mind, are already *abstract* ideas, insofar as they are thus detached from the qualities that exist 'united and blended' in the substances themselves. Indeed, in chapter II. xii, Locke clearly defines abstraction as the act of 'separating [simple or complex ideas] from all other *Ideas* that accompany them in their real existence'. It is this separation that allows the mind to produce 'all its general ideas'. But this definition also suggests that abstraction accomplishes something more fundamental: for example, before it produces the *general idea* of red, it allows the *simple idea* of red to exist as such in the understanding, by separating from it all other ideas that accompany it in 'real existence', as these other ideas could divide attention. Another passage from Book IV confirms this interpretation: 'the proper Operation of the Mind, in Abstraction, [is] to consider an *Idea* under no other Existence, but what it has in the Understanding' (IV. ix. 1). It is clear that the existence of simple ideas is of that kind. Even if the mind is mostly 'passive' in their reception, they cannot exist in it without a minimal activity of perception that make them abstract in this more fundamental sense.

How does this description differ from Descartes and his doctrine on existence? In II. ii. 1, Locke clearly hints at a famous passage of the Second Meditation where Descartes deals with the nature of a piece of wax. The differences between the two accounts are again quite significant. Here is what Locke writes:

> For though the Sight and Touch often take in from the same Object, at the same time, different *Ideas*; as a Man sees at once Motion and Colour; the Hand feels Softness and Warmth in the same piece of Wax: Yet the simple *Ideas* thus united in the same Subject, are as perfectly distinct, as those that come in by different Senses.... And there is nothing can be plainer to a Man, than the clear and distinct Perception he has of those simple *Ideas*. (E II. ii. 1)

And here Descartes:

> But what is this wax which is perceived by the mind alone? It is of course the same wax which I see, which I touch, which I picture in my imagination, in short the same wax which I thought it to be from the start. And yet, and here is the point, the perception I have of it is a case not of vision or touch or imagination—nor has it ever been, despite previous appearances—but of purely mental scrutiny; and this can be imperfect and confused, as it was before, or clear and distinct as it is now, depending on how carefully I concentrate on what the wax consists in.[3]

In this text, Descartes wants to show that the sensory ideas that we have of the piece of wax are merely accidental to it and do not give access to its true nature. This nature consists only in extension in three dimensions, what Descartes calls the 'principal attribute' of the material substance. Descartes wants to reach substantiality beyond sensible ideas. This is the reason why he puts such an emphasis on the *sameness* of the piece of wax through all its transformations. According to him, this substantial identity can only be reached by an '*inspectio mentis*', an inspection of the mind that is entirely different from sight, touch, and imagination, and consists only in a clear and distinct perception of the principal attribute. As the mind finds in itself this idea of

[3] AT 7, pp. 27–8: CSM 2, p. 21.

extension, it could be considered innate. It was already a central claim of the *Regulae*: 'everyone can mentally intuit that he exists, that he is thinking, that a triangle is bounded by just three lines, and a sphere by a single surface, and the like' (Rule 3, AT 10, p. 386; CSM 1, p. 14). Thus, the knowledge of extension is as perfectly clear and distinct as the knowledge of the mind's existence.

When Locke recurs to the Cartesian example of the piece of wax, his intentions are quite different. He is not looking for the principal or essential attributes of bodies. It is only later on, in chapter II. viii, that he will deal with the ideas of primary qualities. At this stage, he does not distinguish between more or less 'substantial' qualities (i.e. 'primary' or secondary), but forcefully insists on the *difference* between existence in a substance and existence of ideas in the mind. If there is here such a thing as an *inspectio mentis*, it consists in the abstract perception that separates various simple ideas from 'real existence' and substantiality. The 'clear and distinct perception', here, applies to the various simple ideas produced in the mind through sensation, and not to what makes a being the same being in its various sensory manifestations.

3. Knowledge of Ideas and Knowledge of Substances

One may wonder why it is for Locke so important to make a distinction between the existence of ideas and 'real' or substantial existence? As we have shown in our discussion of innate ideas, Locke wanted to make clear that the existence of ideas is strictly dependent on the *activity* of perception, and does not involve any postulation of substantial existence. It is clear from several passages of the *Essay* that Locke found that Descartes's doctrine of substance and principal attributes is severely flawed. Here is, for instance, how he criticizes Descartes's identification of space and body:

> He that with *Des-Cartes*, shall frame in his Mind an *Idea* of what he calls *Body*, to be nothing but Extension, may easily demonstrate, that there is no *Vacuum*; *i.e.* no Space void of Body, by this Maxim, *What is, is*. For the *Idea* to which he annexes the name *Body*, being bare Extension, his Knowledge, that Space cannot be without Body, is certain. (E IV. vii. 12)

What is at stake here is not just a quarrel about names. Descartes had indeed a metaphysical argument[4] for the identification of extension and body, namely that any particular extension, being endowed with actual properties, such as a determinate length, breadth, and depth, cannot be nothing, and must therefore be considered the attribute of some substance. The argument is clearly stated in *Principles* 2.16, where Descartes explains why the *vacuum* is not possible:

> The impossibility of a vacuum, in the philosophical sense of that in which there is no substance whatsoever, is clear from the fact that there is no difference between the extension of a space, or internal place, and the extension of a body. For a body's being extended in length, breadth and

[4] On this, see in this volume James Hill's and Lisa Downing's chapters.

depth in itself warrants the conclusion that it is a substance, since it is a complete contradiction that a particular extension should belong to nothing; and the same conclusion must be drawn with respect to a space that is supposed to be a vacuum, namely that since there is extension in it, there must necessarily be substance in it as well.[5]

We see here how Descartes concludes from extension to substantiality. The strategic step in the argument is the impossibility that nothingness should have extension, or properties in general. And this is precisely the principle that Locke rewrites as: 'What is, is.' Descartes would name it a 'common notion', that is an innate principle. But, to Locke, it is the typical example of these trifling propositions 'which under the name of *Maxims* and *Axioms*, have passed for Principles of Science' (IV. vii. 1) but whose self-evidence cannot substantiate any claim on '*real Existence*' (IV. vii. 7). Furthermore, this reference to substance creates a confusion between two very different ideas (the idea of space, and the idea of body) that should be distinguished in a truly clear and distinct perception.

A few years before the *Essay*, a similar issue had been already raised in the quarrel on 'true and false ideas' which opposed Antoine Arnauld and Nicolas Malebranche. Arnauld had strongly criticized Malebranche's claims that ideas are 'representative beings', and that we see them 'in God'. Malebranche's position was based on the very principle that Descartes is making use of for proving the substantiality of extension, that is the maxim saying that 'nothingness has no property':

[A]s for the idea that necessarily exists, and that cannot be other than as it is seen, [men] ordinarily judge unreflectingly that it is nothing—as if ideas did not have a great number of properties; as if the idea of a square, for example, were not different from that of a circle or a number, and did not represent completely different things, which can never be the case for nonbeing, since nonbeing has no properties. It is therefore indubitable that ideas have a very real existence.[6]

The last sentence shows clearly where the problem lies: what sort of existence should we attribute to ideas? According to Malebranche, 'ideas have a very real existence': ideas are real beings, and necessarily partake of substantiality. This is where Arnauld disagrees. According to him, the existence of ideas cannot be separated from their conscious perception in the mind. This is what he calls their 'objective presence',—a term which is contrasted with the 'objective reality' of ideas, in a significant departure from both Malebranche and Descartes. Whether Locke was influenced by Arnauld is a disputed question amongst commentators.[7] It is clear, however, that they both hold the same kind of position about the existence of ideas. They both deny that

[5] P 2.16, AT 8A, p. 49: CSM 1, pp. 229–30. See also Descartes to Arnauld, 29 July 1648 (AT 5, p. 223: '*non satis consideremus nihili nullas proprietates esse posse*').

[6] SAT 3. 2. 1, trans. in LO.

[7] For arguments in favour of this influence, see Yolton, *A Locke Dictionary*, article 'Idea', p. 91. In a passage of The Epistle to the Reader added to the fourth edition of the *Essay*, Locke defines determinate ideas as ideas that are 'objectively in the Mind' E Epistle, p. 13.

ideas could consist in substantial beings; both construe this existence as an actual perception in the mind.[8]

One central claim of the *Essay*, especially in Book IV, is that human knowledge first deals with ideas, rather than substances. This is what John Yolton called Locke's 'way of ideas'[9]. Actually, it may be more accurate to think of it as a necessary *detour*: all our knowledge should eventually terminate in real beings and substance; however, the first objects on which we apply our faculties are always ideas. That is why Locke raises in chapter IV. iv the question of the 'Reality of our Knowledge': he tries to prevent his reader from thinking of his doctrine as the construction of 'Castles in the Air'; and he intends to show that 'this way of certainty, by the Knowledge of our own *Ideas*, goes a little farther than bare Imagination' (IV. iv. 2).

The first argument here is that our simple ideas, insofar as they cannot be produced by the human mind, 'are not fictions of our fancies'. As finite creatures, we have to rely on the belief that they have a natural and regular correspondence with things outside, and the operations they exert upon our mind. According to Locke, 'this conformity between our simple ideas, and the existence of things, is sufficient for real knowledge'. Again, we can see that Locke distinguishes not only ideas from things, but also two different kinds of existence.

The second argument is that all complex ideas, except those of substances, can be considered '*archetypes* of the mind's own making', and therefore are their own measure of reality and truth, regardless of any substantial supposition. As a consequence, there is no need to refer them 'to the existence of any thing'. Acknowledging the independence of ideas, and their use by the mind, opens up a whole field of knowledge, where the consideration of real existence, or substances, would be not only useless, but also an obstacle:

> we may take notice that *universal Propositions* of whose Truth or Falsehood we can have certain knowledge concern not *Existence*: and further, that all *particular Affirmations or Negations*, that would not be certain if they were made general, are only concerning *Existence*; they declaring only the accidental union or separation of *Ideas* in Things existing, which, in their abstract Natures, have no known necessary Union or Repugnancy.[10]

Here, Locke shows that we can always make a crucial choice in the use we make of our own ideas. On the one hand, we can try to seize their 'union or separation in things existing' (what he also calls their 'coexistence' in substances); but this kind of knowledge will be mostly particular and accidental, and 'very short.' In most cases there will be no necessity in the connection of substantial qualities. On the other hand, our intuitive or demonstrative knowledge of the 'relations of ideas' can reach certainty and make spectacular progress, as soon as we set aside any consideration of real existence. This text clearly shows that the intellectual world of ideas must be to some extent

[8] Descartes's use of the term 'idea' was already ambiguous: sometimes it means the object of a mental act; sometimes it means the mental act itself. On this ambiguity, see Jolley, *The Light of the Soul*, ch. 2.

[9] Yolton, *John Locke and the Way of Ideas*.

[10] E IV. ix. 1.

disconnected from the world of real existence. Of course, it is particularly true in mathematics. But it also includes moral matters, as they are for Locke capable of demonstration:

> All the Discourses of the Mathematicians about the squaring of a Circle, conick Sections, or any other part of Mathematicks, *concern not* the *Existence* of any of those Figures: but their Demonstrations, which depend on their *Ideas*, are the same, whether there be any Square or Circle existing in the World, or no. In the same manner, the Truth and Certainty of moral Discourses abstracts from the Lives of Men, and the Existence of those Virtues in the World, whereof they treat.[11]

Of course, all these demonstrations may eventually produce some valuable knowledge about things in real existence, insofar as the ideas agree with these archetypes: for example, what we know about the idea of a triangle can be applied to a material triangle in real existence. But it makes it all the more necessary to deal first with these ideas considered in themselves, in bare perception, because all these general truths cannot be found in the consideration of material substances. Therefore, when a mathematician deals with geometrical figures, 'he is sure what he knows about those Figures, when they have barely *an Ideal Existence* in his Mind, will hold true of them also, when they have a real existence in Matter; his consideration being barely of those Figures, which are the same, where-ever, or however they exist' (IV. iv. 6). This sentence reflects Locke's implicit doctrine about existence—a doctrine that is clearly dualistic: there is always a gap between ideal existence (in the mind) and real existence (referred here to matter), even if they may agree or correspond to some extent. This dualism does not consist in distinguishing two kinds of substances (material and spiritual), but two kinds of *existence*—one of which (the existence of ideas in the mind) is clearly and deliberately deprived of any substantiality.

Locke's doctrine of 'modes' is a good illustration of how this ontological claim of the *Essay* arose out of critical reflection on Descartes's philosophy. In Descartes's vocabulary, modes are referred to substances; in the *Essay*, Locke deals with the modes—or modifications—of *ideas*. But the striking point here is the way Locke explains how he was led to *transfer* this notion from the (Cartesian) world of substances to the world of ideas.

According to Descartes, modes are 'attributes' or 'qualities' of substances; but, unlike extension and thought that are the 'principal attributes' of the two kinds of substance and constitute their nature, modes are the non-essential features of substances, such as the particular size and shape of a body, or the particular thoughts of a mind; they are therefore the features that are submitted to change or can be modified, or the contingent part of substances. This is why Descartes says that God, in his perfection, cannot have 'modes', but only 'attributes'.[12]

Now, in the *Principles*, the relationship between principal attributes and modes raises a major problem: the principal attribute 'constitutes the nature' or the essence of

[11] E IV. iv. 8.   [12] P 1.56, AT 8A, p. 26: CSM 1, p. 211.

the substance; but it is also what allows modifications of the same substance—and therefore its modes—to take place. In other words, the feature that does not and cannot change in a substance—extension for bodies, thought for minds—is also, but from another point of view, the feature that must welcome every change of the same substance. Extension, for example, is the identical nature of any material substance, but must at the same time take different shapes, positions, etc. More importantly, the modal changes encountered in experience may lead us to believe that the principal attribute itself can change in the substance. For example, when a body moves and leaves its position in space, we tend to believe that it has also left *its extension* there. This is why Descartes insists that we should never forget that the principal attribute cannot exist by itself, and must always be considered the attribute *of a substance*:

> Thought and extension can be regarded as constituting the natures of intelligent substance and corporeal substance....
>
> Thought and extension may also be taken as modes of a substance, in so far as one and the same mind is capable of having many different thoughts; and one and the same body, with its quantity unchanged, may be extended in many different ways...The distinction between thought or extension and the substance will then be a modal one; and our understanding of them will be capable of being just as clear and distinct as our understanding of the substance itself, provided they are regarded not as substances (that is, things which are separate from other things) but simply as modes of things.[13]

According to this text, the difficulty arises from a defect of the understanding, and more precisely from a lack of *attention*, that makes us prone to believe that thought and extension could subsist by themselves, even when there is no substance to 'support'[14] them. The last sentence clearly shows Descartes's tendency to seek substantiality beneath the various qualities we find in experience.

Now, when Locke gives a first definition of the 'ideas of modes', it seems quite clear that he has this Cartesian problem in mind. The first sentence of his definition is purely and simply the translation of Descartes's last sentence in the previous quotation:[15]

> First, *Modes* I call such complex *Ideas,* which however compounded, contain not in them the supposition of subsisting by themselves, but are considered as Dependences on, or Affections of Substances;—such are the *Ideas* signified by the Words *Triangle, Gratitude, Murther, etc.* And if in this I use the word *Mode,* in somewhat a different sense from its ordinary signification, I beg pardon; it being unavoidable in Discourses, differing from the ordinary received Notions, either to make new Words, or to use old Words in somewhat a new signification, the latter whereof, in our present case, is perhaps the more tolerable of the two.[16]

[13] P 1.63–4, AT 8A, p. 31: CSM 1 p. 215. In the French translation of Abbé Picot the last sentence is slightly different: '...pourvu que nous ne pensions point qu'elles subsistent d'elles-mêmes mais qu'elles sont seulement des façons ou dépendances de quelque substances' ('provided that we do not think that they subsist by themselves, but consider that they are only the manners or dependences of some substance').

[14] For Locke's emphasis on this term, see II. xiii. 19.

[15] Or rather a translation of the French version of this passage (see note 13).

[16] E II. xii. 4.

Even if Descartes is not mentioned by name, it is quite plausible that Locke's justification of his own reinterpretation of the 'received notion' comes from a critical reading of §64 of the first part of the *Principles*. What is at stake here is a 'new signification' of the term 'modes'; but ironically, Locke found it in Descartes's own formulations. Locke actually agrees with Descartes on one point: modes 'contain not in them the supposition of subsisting by themselves'. But he gives to this characterization an anti-Cartesian twist. In the *Essay*, modes will be indeed considered without any 'supposition' (II. xii. 4) of substantiality. According to Locke, they are the result of an intellectual activity, that of the understanding compounding and modifying its own simple ideas. The Cartesian urge to look for substances everywhere can only hinder the liberty of framing new ideas in this intellectual world—even if, eventually, all these ideas shall be referred to things actually existing without us. Locke says it at the beginning of chapter II. xiii: simple modes are

> *Modifications of the same* Idea; which the Mind either finds in things existing, or is able to make within it self, without the help of any extrinsical Object, or any foreign Suggestion. (II. xiii. 1)

The same can be said of mixed modes:

> [the mind] being once furnished with simple *Ideas*, it can put them together in several Compositions, and so make a variety of complex *Ideas*, without examining whether they exist so together in Nature. And hence, I think, it is, that these *Ideas* are called *Notions*: as if they had their Original, and constant Existence, more in the Thoughts of Men, than in the reality of things. (II. xxii. 2)

One can see here how Locke not only criticizes Descartes's ontology, but takes advantage of Cartesian concepts to give consistency to his own intellectual world, which is based on the existence of ideas and operations rather than substances—and where ideas of modes are also modes of ideas...

4. The 'Virtual Ontology' of the *Essay*

How consistent and autonomous is the intellectual world? The question raises a number of significant difficulties. Locke doesn't elude them: on the contrary, his efforts to answer them result in what we propose to call the 'virtual ontology' of the *Essay*. In this section, we shall try to give a general idea of what it may consist in.

Despite all the efforts that Locke makes to consider an idea 'under no other existence, but what it has in the understanding', that is, without any substantial supposition, it is clear that the two kinds of existences must eventually coincide at some point. Locke's intellectual world, based on perception, must be connected in some way to the 'real existence' of substances—of things that 'exist' *even if (and when) we do not actually perceive them.*

This connection is especially needed. The perception of ideas being a transient and changing operation, it creates a transient and changing intellectual world. This is perhaps one of the most puzzling aspects of Locke's doctrine: when an idea is the object of an actual perception, and therefore exists in the understanding, it cannot be there for long, because sensation and reflection necessarily 'push' other ideas into the mind. It is a question whether the *same idea* can appear twice in the understanding. Locke's answer is clear: each particular idea is a new one, the new mental event is never the same as before. Even if we try and intently focus on an idea, it will be necessarily altered in the continued operation of perception:

> For Trial, let him take any Figure, any Degree of Light or Whiteness, or what other he pleases; and he will, I suppose, find it difficult to keep all other *Ideas* out of his Mind: But that some, either of another kind, or various Consideration of that *Idea* (each of which Considerations is a new *Idea*), will constantly succeed one another in his Thoughts, let him be as wary as he can. (E II. xiv. 14)

So even if we try hard to keep one idea in our mind, the various considerations of this idea are as many new and different ideas. Similarly there is no one idea of memory that has exactly the same content as the previous idea it recalls, as there will be at least 'this additional Perception annexed to [it]', that the mind has had it before (II. x. 2). Of course, Descartes or Malebranche would say that the 'objective reality' of the idea—for example the objective properties of a triangle—must remain the same, even if we do not actually consider it. But we have already seen that Locke rejects any intellectual reality that would not depend on an actual perception. Therefore, the existence of an intellectual world of ideas—especially of modes such as geometrical figures, and demonstration about them—cannot mean that these ideas exist somewhere 'above' or 'beyond' our momentary operations of perceptions. It would be much more accurate to say that these ideas we would like so much to be the same in the various considerations we have of them, have actually no existence at all. It would be a possible—although radical—way of understanding a passage already quoted: if the demonstrations of the mathematicians 'concern not existence', it is because the ideas they deal with have no actual existence. One could say that they only have a *virtual* existence—which is still something. That's why Locke puts so much stress on the need for 'visible and lasting marks, wherein the *Ideas* under consideration are perfectly determined' (E IV. ii. 10). It is the perception of these marks that actually exists in the understanding, rather than the ideas themselves. At the very end of the *Essay*, Locke even comes to consider that ideas in general are nothing but signs of 'things' that cannot exist actually in the mind: 'since the Things, the Mind contemplates, are none of them, besides it self, present to the Understanding, 'tis necessary that something else, as a Sign or Representation of the thing it considers, should be present to it: And these are *Ideas*' (E IV. xxi. 4).

It is for the same kind of reason that Locke distinguishes in Book IV between 'actual' and 'habitual' knowledge. Given the limitations of our actual perception, our knowledge would not go very far if we couldn't rely on the 'habitudes' of ideas. For instance, we rely on a known theorem in a demonstration, without actually verifying its truth in an actual perception:

If then the Perception that the same *Ideas* will eternally have the same Habitudes and Relations be not a sufficient ground of Knowledge, there could be no knowledge of general Propositions in Mathematicks, for no mathematical Demonstration would be any other than particular.

(E IV. i. 9)

We can see here to what extent the intellectual world of the understanding has most of the time a *potential* or *virtual* existence. The stress that Locke puts on habitudes, and more generally on the distinction between *potential* and *actual* perception, shows that his world of ideas owes much more to Aristotle than to Plato.

This is also why the idea of *power* is so important to Locke. The transient and changing existence of ideas in actual perception leads directly to it:

The Mind, being every day informed, by the Senses, of the alteration of those simple *Ideas*, it observes in things without; and taking notice how one comes to an end, and ceases to be, and another begins to exist, which was not before; reflecting also on what passes within it self, and observing a constant change of its *Ideas*,...; and concluding from what it has so constantly observed to have been, that the like Changes will for the future be made, in the same things, by like Agents, and by the like ways, considers in one thing the possibility of having any of its simple *Ideas* changed, and in another the possibility of making that change; and so comes by that *Idea* which we call *Power*. (E II. xxi. 1)

We can see here the origin of our need to seek and suppose substantiality beneath the constant changes of our ideas: we have to imagine that these changes originate from the operations of the 'same things'. Therefore, we can understand why the existence of ideas must depend eventually on some *real* existence that explains it. However, according to Locke, we cannot reach substances (even our own 'thinking substance') but only the *powers* they exert on our ideas.

Now, amongst these various powers that determine the very possibility of our intellectual world, *sensation* plays a central role. It is the first faculty that gives the mind some matter, something to think on. It is therefore the very first step between potential and actual perception, that initiates the use of all our intellectual faculties:

Follow a *Child* from its Birth, and observe the alterations that time makes... [T]here appear not to be any *Ideas* in the Mind, before the Senses have conveyed any in....'Tis about these Impressions made on our Senses by outward objects, that the Mind seems first to employ it self in such Operations as we call *Perception*, *Remembring*, *Consideration*, *Reasoning*, etc.

(E II. i. 22–3)

There is another reason why sensation is so important to Locke. Whereas reflection gives ideas of 'operations', sensation provides the mind with the only ideas that are apt to give consistency to the world of perception: ideas of primary qualities (extension, solidity, figure, and mobility). Of course, according to Locke, these qualities are also 'powers' (E II. viii. 8). Nevertheless, their constant presence in experience gives coherence to the world of our perception. This is particularly true of solidity:

The *Idea* of *Solidity* we receive by our Touch; and it arises from the resistance which we find in Body, to the entrance of any other Body into the Place it possesses, till it has left it. There is no

> *Idea*, which we receive more constantly from Sensation, than *Solidity*. Whether we move, or rest, in what Posture soever we are, we always feel something under us, that supports us, and hinders our farther sinking downwards; and the Bodies which we daily handle, make us perceive, that whilst they remain between them, they do by an insurmountable Force, hinder the approach of the parts of our Hands that press them (E II. iv. 1)

We always feel something under us, that supports us: the idea of solidity, such as provided by sensation, is the most 'substantial' of our ideas. But what exactly does solidity 'support'? The passage suggests that it is not only our body, but our whole being, which also includes our thoughts. Indeed, it may be only through sensation and solidity that the transient and mostly virtual existence of ideas in the mind can be eventually based on some *real* existence. It would explain why chapter II. vii distinguishes between reflection that gives us access to the (virtual) existence of ideas, and sensation, in which only we have some access to the existence of 'things'.

Since the intellectual world of ideas cannot subsist by itself, it depends on the world of 'real existence'. It happens only through sensation (through solidity, but also through the 'sensible marks' the mathematicians use in their demonstrations, for example). Descartes said that we can transfer our idea of existence from our mind to bodies. Locke would agree to the possibility and even to the necessity of a transfer. But, according to him, it goes the other way round: we transfer, in metaphors, our ideas of sensation, and more particularly our idea of solidity, from the material world to the intellectual world of reflection.[17] Without this transfer, we couldn't give any consistence to the virtual existence of ideas. In the *Essay*, the problem of existence is always a problem of *double meaning*: there is a gap between the existence of 'real things' (such as bodies) and the existence of ideas; but at the same time these two meanings of existence must be connected in some way, at least through metaphors, because they both partake of one human reality. We can see this, for example, in the description of memory:

> *Memory*... is as it were the Store-house of our *Ideas*.... But our *Ideas* being nothing, but actual Perceptions in the Mind, which cease to be any thing, when there is no perception of them, this *laying up* of our *Ideas* in the Repository of the Memory, signifies no more but this, that the Mind has a Power, in many cases, to revive Perceptions, which it has once had, with this additional Perception annexed to them, that it has had them before. And in this Sense it is, that our *Ideas* are said to be in our Memories, <u>when indeed, they are actually no where</u>, but only there is an ability in the Mind, when it will, to revive them again. (E II. x. 2; my emphasis)

In memory, ideas only have a potential or virtual existence, insofar as they *can* be revived in an actual perception. It means that memory is a power of the mind. The best way to describe it in reflection is to use the metaphor of a 'Store-house', a place that would support this virtual existence of ideas, even when they are not actually perceived. There is no substantial supposition here, but only a 'floating vision' of the

[17] See E III. i. 5, where Locke describes how '*obvious sensible* Ideas *are transferred to more abstruse significations*, and made to stand for *Ideas* that come not under the cognizance of our senses'.

unknowable thinking thing that supports the operations of perception. Actually, Locke often suggests that metaphors of 'solidity' are an essential part of our need for substantiality. This is why he insists on the etymology of the word *substantia*, which means 'standing under'.

Thus, the status of the existence of ideas is quite elusive in the *Essay*, especially when considered in its relationship with 'real existence'. On the one hand, ideas exist in the ideal or intellectual world of perception, where they must be independent of any substantial 'supposition'. One of the most striking aspects of Locke's ontology is that it makes much room for such an 'ideal' existence, against Descartes's ontology of substance. It explains for example why modes are in the *Essay* modes of ideas rather than modes of substances. It also explains Locke's conception of a 'clear and distinct' perception, in which the mind perceives simple ideas, rather than 'principal attributes'.

On the other hand, it is clear that this ideal world of ideas cannot *subsist* by itself. It is a virtual world, mostly made of potential operations of perceptions. As such, it must depend on a 'real existence' that 'supports' it. According to Locke, since we cannot reach substances, this support in real existence consists in *powers* and especially in *sensation*, which is the main thing that gives consistency to our train of perceptions. Locke's ontology is therefore split between two kinds of existence. That is why Locke often uses double meaning when it comes to their relationships. Metaphors are therefore an essential feature of what we may call the virtual ontology of the *Essay*. Hannah Arendt, in *The Life of the Mind*, has shown to what extent metaphor can be much more than a stylistic device: the only way we can establish a link between the world of sensible things and the invisible and non-phenomenal world of our own thought: 'The metaphor, bridging the abyss between inward and invisible mental activities and the world of appearances, was certainly the greatest gift language could bestow on thinking and hence on philosophy.'[18] Locke is definitely a gifted philosopher in this respect. He was certainly well aware of the possibilities offered by language and metaphors: it is not surprising for example that he should have translated the Latin word '*substantia*' into the English phrase 'standing under'—in such a way we can see how close it is to the word 'understanding'.

References

Arendt, Hannah. *The Life of the Mind*. New York: Harcourt Brace & Company, 1977.

Jolley, N. *The Light of the Soul: Theories of Ideas in Leibniz, Malebranche, and Descartes*. Oxford: Oxford University Press, 1990.

Yolton, John W. *John Locke and the Way of Ideas*. Oxford: Oxford University Press, 1956.

Yolton, John W. *A Locke Dictionary*. London: Blackwell, 1993.

[18] Arendt, *The Life of the Mind*, p. 105.

7

Locke and Descartes on Selves and Thinking Substances

Philippe Hamou

Locke's account of selves, persons, and thinking substances is notoriously difficult and the subject of wide controversy. In this chapter, I propose to go some way towards clarifying it by viewing it in the context of Descartes's approach to the same or similar objects.

Three typical questions may be raised concerning selves and thinking substances in Descartes and Locke. The first is whether it is true, as Descartes contends in the *Meditations*, that we can gain a clear and distinct knowledge of the *nature* of the thinking thing that we are. The second question, closely related, is whether this nature is that of a thing (a substance) whose whole essence is to think. Locke, already in the first edition of the *Essay* (that is, before the introduction of the chapter on Identity and Diversity in the second edition), argues against Descartes's positive answers to both questions, defending the view that we do not have clear and distinct knowledge of the thinking substance, but can nevertheless rule out the suggestion that thought is the 'essence' of it. One may consider this polemical stance as the characteristically anti-Cartesian backbone of the *Essay*, insofar as the question of the nature of the soul is concerned. The third question is whether being the same *self* necessarily entails being the same (immaterial) substance. It is of course one of the central issues of chapter II. xxvii on identity. But here the comparison with Descartes's positions is less straightforward. The issue of personal identity is not one that Descartes directly addresses. Besides, although Locke's treatment of the question still contains hints against Cartesianism, there are reasons to think that, in introducing a new concept of 'person' for designating the subsisting self, and in making it entirely independent from the persisting conditions of any sort of substratum, Locke was here perhaps closer to the true spirit of Cartesian philosophy than he ever was—provided that one is content to relinquish the usual caricature of Cartesian dualism and, following recent reappraisals, assigns to Descartes a richer and more nuanced view of what it takes to be an immaterial mental substance.

1. Conscious Thinking

One shared feature of Locke's and Descartes's approaches to these questions is their common commitment to the thesis according to which *consciousness* is inseparable of the very experience of thinking, or, as Locke put it, 'essential to it' (II. xxvii. 9). Locke said repeatedly that we cannot think, 'have ideas', without being conscious of them, that is, without conceiving at the same time, being firmly convinced that these thoughts and/or ideas are *ours*. 'It is impossible for anyone to perceive, without perceiving, that he does perceive.' (II. xxvii. 9) These formulas are not simply suggesting the weak implication: *if x is a thought of ours, it is always possible to be conscious of x*; but quite clearly point to the strong one: *if x is a thought of ours, it is necessary that we are conscious of x*. The formulas however do not mean that *being conscious of having ideas* is the same as *having ideas of this or that object*; rather what they suggest is that these are same order phenomena occurring concomitantly in one's mind. We shall come back to this later.

Locke's terms may be compared with important passages from Descartes, such as this one from the answers to Arnauld's Fourth set of Objections: 'we cannot have any thought of which we are not aware at the very moment when it is in us' (*nec ulla potest in nobis cogitatio, cujus eodem illo momento, quo in nobis esse conscii non simus*).[1] The formulations are strikingly similar. Descartes, just like Locke, is not saying that having a thought is the same as being conscious, but rather that consciousness always accompanies thought: while we are thinking such or such *objects*, we are conscious of our *thoughts*.[2] Descartes sometimes says that consciousness is the defining criterion through which we know what sort of things in us are thoughts. In the second replies, he writes: 'Thought (*cogitationis*): I use this term to include everything that is within us in such a way that we are immediately aware (*immediate conscii*) of it.'[3] To Arnauld, who suggested that our mind may have ideas or faculties of which we are not conscious, he answers that we still have 'potential consciousness' of them, explaining that if such un-conscious faculty is in us, we should be able to summon it at will: wherever it is, we cannot but be aware of it as soon as we prepare ourselves to use it.[4] This clearly indicates that Cartesian consciousness describes the sort of native familiarity that

[1] AT 7, p. 246; CSM 2, p. 171. The phrase is taken verbatim from Arnauld's objections, where it is used in order to account for the passage of the Third Meditation where Descartes suggests that if any power is supposed in my mind, I should be conscious (*conscius*) of it. (AT 7, p. 49; CSM 2, p. 234). It has been noted (Alanen, 'Thought') that this passage is a unique occurrence in the whole text of the Meditations, where the substantive 'consciousness' (*conscientia*) is totally absent. We should point out however that Descartes uses 'conscius essem' there, quite significantly, as an equivalent for 'experior'. Adopting Arnauld's slogan as his own, he later consistently uses 'conscius esse' in reference to the inner experience that we cannot but have of our own thinking. Thanks to Martine Pécharman for bringing this to my attention.

[2] On Descartes's understanding of consciousness see especially Radner, 'Thought and Consciousness', and Simmons 'Cartesian Consciousness Reconsidered'.

[3] AT 7, p. 160: CSM 2, p. 113. See also P 1, p. 9 which defines thought as 'everything which we are aware of (nobis consciis) as happening within us, in so far as we have awareness of it' (AT 8A, p. 7: CSM l, p. 195).

[4] AT 7, pp. 246–7: CSM 2, p. 172.

we have with our own mind. It seems that it is just as when we use our limbs, hands, or eyes, we need not be instructed in order to find our ideas or mental faculties where they are, or to know how we should use them.

Despite these similitudes, Étienne Balibar[5] has recently argued that Locke 'invented' a modern idea of consciousness, precisely in contrast to Descartes's uses of '*conscientia*'. To him, consciousness, in the 'Lockean', modern, sense, is an introspective activity, employed to explore the mental scene, the contents of the mind and its various operations. Such *psychological* programme, he argues, was never intended by Descartes, whose *conscientia* accompanies all thought simply because it expresses the 'self-presence of thought' ('*la présence à soi de la pensée*'), which 'does not depend on the exercise of any faculty'.[6]

What is debatable in this presentation is that *consciousness* and *reflection* are used as almost interchangeable terms. Balibar presents the former in the very terms that Locke uses to construe the latter[7]—that is, as an 'internal sense' through which the mind acknowledges its own faculties, operations, and ideas.[8] This quasi-synonymy may be nowadays more or less implicit in a widespread notion of consciousness as a scanning faculty through which we get 'access' to certain inner states of our minds,[9] but I do not think it is right to assign it to Locke. Reflection is described in the *Essay* either as a receptive faculty that gives ideas of our own mental operations, or as an introspective attention that can be oriented willingly on such or such parts of our minds.[10] Consciousness by contrast is never presented as a potential source of 'new' ideas. It is not a receptive faculty, but neither is it a voluntary action of the mind that we could choose to exert or not. One could say that it is not a faculty at all, and indeed, it is never included in the various inventories of our mental operations that one finds in chapter II. I, or II. xi. It is more properly described as a constitutive and common feature of all thoughts and mental operations whatsoever.[11] It is true that Locke sometimes defines reflection and consciousness in terms that look similar. 'Consciousness

[5] Balibar, in his *Identité et difference*, whose subtitle is 'l'invention de la conscience'.

[6] *Identité et différence*, p. 41.

[7] For example, p. 26: '*elle est une reconnaissance immédiate par le mind de ses opérations sur la "scène" intérieure*'; again, p. 38, talking of Descartes's *conscientia* in its alleged contrast with Locke's consciousness: 'il n'est donc pas question de fixer à cette conscience un programme d'investigation réflexive tel que découvrir les facultés de l'âme ou analyser des opérations logiques . . . contrairement à ce que croiront certains successeurs de Descartes.'

[8] Cf. II. i. 4: 'By reflection then, in the following part of this discourse, I would be understood to mean, that notice which the mind takes of its own operations, and the manner of them, by reason whereof there come to be ideas of these operations in the understanding.'

[9] David Armstrong defends this view of consciousness as scanning of the inner states of the mind, and ascribes its origin to Locke. See *A Materialist Theory of the Mind*, p. 198. On the distinction between phenomenal and access consciousness, see Chalmers, *The Conscious Mind*.

[10] On this dual sense of reflection see Lähteenmäki, 'The Sphere of Experience'.

[11] See II. i. 4, which subtly articulates reflection, actings of the mind, and consciousness, without confusing them: reflection 'furnishes the Understanding with another set of ideas . . . : and such are perception, thinking, doubting, believing, reasoning, knowing, willing, and all the different actings of our own minds;—which we being conscious of, and observing in ourselves, do from these receive into our understandings as distinct ideas as we do from bodies affecting our senses'.

is the perception of what passes in one Man's own mind' (E II. i. 19); reflection 'is the *Perception of the Operations of our own Minds* within us' (E II. i. 4). However close they look, these are different definitions, because the term perception is taken in two different senses. It is one thing to perceive (to be aware) of what passes in our mind, and quite another to perceive (observe) the operations of our mind when it perceives (or is conscious of) whatever passes in it. The perception involved in consciousness, rather than a higher order act, is more correctly described as a first order phenomenon, a 'built in', self-representational character of our thoughts.[12] Our thoughts *present* themselves to us, in the very act of representing their object. Their being conscious is the way our thoughts *feel* to us, and this is why it is an inner and immediate characteristic of the thoughts themselves, 'essential' to them in this sense. It belongs to the way these thought are 'experienced', 'intimated', or intimately 'felt', rather than *reflected*, or known.

Thus, contrary to Balibar, I tend to think that Locke was a rather faithful Cartesian on the topic of consciousness. His acceptation of the term in the *Essay* is not fundamentally different from the one that became usual among the French Cartesians such as Arnauld or La Forge, authors whom Locke read when he was in France, and who both explicitly distinguished between consciousness and reflection. La Forge in particular, gives us what is perhaps the clearest account of the distinction:

> I think I can define the nature of thought as that consciousness, awareness and inner feeling by which the mind is aware of everything it does or suffers, and in general, of everything which takes place immediately in itself at the same time as it acts or is acted on. I say 'immediately' to let you know that this testimony and inner feeling is not something different from the action or passion and that the actions and passions themselves make the mind aware of what is taking place in itself. Thus you will not confuse this inner feeling with the reflection that we sometimes make on our actions, which is not found in all our thoughts because it is only one type of thought.[13]

It is interesting to observe that Locke's French translator, Pierre Coste, lacking a French equivalent for consciousness, embarrassed by the moral connotation of the French word 'conscience', first chose to translate with such phrases as *conviction intérieure*, or *sentiment intérieur*, before opting (but only in chapter II. xxvii) for the typographic idiosyncrasy 'con–science'. This clearly reminds us of La Forge's definition of consciousness as an *inner and immediate feeling*. I think it is also quite an adequate rendering of what Locke really meant: the kind of 'knowledge' or conviction that consciousness gives us is not a proposition about the content of our mind, an 'agreement of ideas'

[12] This view of consciousness is reminiscent of Brentano's concept of inner-perception. The affinity between Locke and Brentano has been suggested by Rabb, *John Locke on Reflection*, and more recently in Coventry and Kriegel, 'Locke on Consciousness'. The contrast between reflection and consciousness is also defended in Thiel, 'The Early Modern Subject'.

[13] La Forge, *Treatise on the Human Mind*, p. 57. See also Arnauld, *Des vraies et des fausses idées*, ch. 24, who defines 'la conscience' as 'la perception que nous avons de notre pensée lorsque nous la connaissons *comme par elle-même, sans réflexion expresse*'.

that could be made public or shared: it is not a piece of reflective knowledge in the technical sense, but rather an internal and inarticulate conviction that the thought is ours—internal, in the sense that it is given, or experienced with the thought itself.[14]

2. Lockean and Cartesian *Cogitos*

In the very brief chapter of Book IV devoted to 'our knowledge of existence',[15] Locke puts to use this *Cartesian* concept of consciousness, in order to show that we have an *intuitive* knowledge of our own existence:

> ... nothing can be more evident to us than our own existence. I think, I reason, I feel pleasure and pain: can any of these be more evident to me than my own existence? If I doubt of all other things, that very doubt makes me perceive my own existence, and will not suffer me to doubt of that. For if I know I feel pain, it is evident I have as certain perception of my own existence, as of the existence of the pain I feel. (IV. ix. 2)

Locke's point here seems to be this: our own existence is *as certain* as the very existence of any one of our thoughts in our minds, because an awareness of ourselves is attached to any one of them. Again, one should note that there is no mention here of reflection. According to Locke's former descriptions in chapter IV. ii, the mere fact that we apprehend our ideas in our minds as being such and such, representing red rather than black, for example, is the object of an *intuitive knowledge*, to which the highest degree of certainty belongs. 'For a Man cannot conceive himself capable of a greater Certainty, than to know that any *Idea* in his Mind is such, as it perceives it to be.' (IV. II. 1). As one essential aspect of any one of our ideas is that it is *ours*, felt within ourselves, awareness of ourselves, what Locke will later call 'self-consciousness',[16] inherits the same quality of certainty and immediacy that we find in the intuitive knowledge that the ideas that we have are such or such. This self-reference is, so to speak, embedded into ideas.

[14] Descartes may have himself suggested the idea of consciousness as 'internal knowledge' in a passage from the *Sextae Responsiones*, I (AT 7, p. 422: CSM 2, p. 285), where he says that we understand what is thought or existence through this 'interna cognitio', which he contrasts with the knowledge that we can gain from reflection ('scientia reflexa'). The distinction however concerns another sort of 'tacit' knowledge, the one we have of the meaning of words such as 'thought' or 'existence'. In other terms, it is the kind of tacit knowledge that we have of *innate ideas* when we do not think of them. On this dispositional conception of innate ideas, see *Notae in programma*, AT 8B, p. 361.

[15] The chapter is the first (and by far the shortest) in a sequence of three chapters of Book IV, where Locke proposes his own version of what can be characterized as a Cartesian itinerary: he shows first that we have an intuitive knowledge of the existence of our selves. He continues then (chapter x) arguing that from what we know of ourselves as thinking intelligent beings we can prove demonstratively God's existence. Finally, he shows (chapter xi), that considering the nature of our sensory ideas, and God's benevolence, the existence of an external world can be asserted with certainty.

[16] This famous locution only appears once in Locke, in a note from his journals of 1696 '*our own existence is known to us by a certainty yet higher than our senses can give us of the existence of the other things, and that internal perception is a self consciousness*' (MS Locke c28, fos. 119^{r}–120^{v}, quoted in King, *Life and Letters*, vol. 2, p. 138).

This may be considered as the Lockean version of the *cogito*.[17] But of course, it presents striking differences from Descartes's original. Whereas Descartes considered the '*sum, existo*' of the second Meditation as the very first truth and the paradigm of clear and distinct truths, the Lockean truth of my own existence is only one among many other intuitive truths, such as the truth that our idea of white is distinct from our idea of black, or the intuition of simple mathematical truths, such as 'Three are equal to One and Two' (IV. ii. 1).[18]

Another difference is the role of universal doubt. In Descartes, although any individual act of thought would be able to convey the truth of my own existence, the *cogito* of the *Meditations* owes a special quality to the fact that it is drawn from one specific (and somewhat monstrous) thought: the thought that everything that I used to hold true is false, even the opinion that I have a body, at present writing and sitting beside the fire, or the opinion that 3 and 2 added together are 5.[19] With such a thought in the mind, there remains only one undeniable fact: that *I am thinking such a thought*. This, if we are to believe Descartes, gives to the actual *cogito* of the *Meditations, an epistemic superiority* over any other way to prove my own existence.

Admittedly, Locke pays tribute to the famous argument of Meditations 1 and 2, when he writes: 'If I doubt of all other Things, that very doubt makes me perceive my own *Existence* and will not suffer me to doubt of that' (IV. ix. 2). Here, however, universal doubt is only an illustration of a point that could be made, perhaps more directly, with any other ideas or conscious experiences. Among them, Locke's privileged experience, the one that makes us the most clearly see that we cannot doubt of ourselves, is not *doubt* but *pain*. This can be seen for example in the rather prosaic rephrasing of the argument in the next chapter:

> If any one pretends to be so skeptical as to deny his own existence (for really to doubt of it, is manifestly impossible) let him for me enjoy his beloved Happiness of being nothing, until Hunger, or some other Pain, convince him of the contrary. (IV. x. 2)

Interestingly 'pain' is also invoked in the next chapter, as the best pragmatic answer to scepticism in regard to the existence of the external world.[20] I take this as a sign that in his '*cogito*' argument, Locke is not so much interested in proving the existence of his

[17] Descartes's name is not mentioned in this chapter of the *Essay*. It was however, with approval, in the corresponding passages from Draft A and B. See Draft A (§§10, 27), B (§35). For example, A §27 (p. 42) 'the Understanding knows undoubte[d]ly that while it thinks, reason and imagins, it is or hath existence. Or that there is something that knows & understands which according to Cartes & I think in truth is the most certain & undoubted proposition that can be in the minde of a man.'

[18] From what can be inferred from a letter of John Wynne (1695), answering a missive from Locke that has not survived, it seems that Locke agreed to distinguish among intuitive truths three types of perceptions: perception of self-evident truths, internal perception of our own existence ('self-consciousness'), and the consciousness of ideas. Locke's '*cogito*' shows how the third and the second are intimately connected. Cf. Wynne to Locke, 1695, LC 5, 347 letter 1884 and Thiel, *The Early Modern Subject*, p. 119.

[19] See Descartes's *Meditatio prima*, AT 7, p. 20: CSM 2, p. 14.

[20] See E IV. xi. 3: 'As to my self, I think GOD has given me assurance enough of the Existence of Things without me: since, by their different application, I can produce in my self both Pleasure and Pain, which is one great Concernment of my present state.'

being as a spiritual, immaterial thing as Descartes is,[21] as his existence as a *man*—that is, a fully embodied thinking being. More clearly than doubt and reason, feelings like pains and pleasures are directly attached to the embodied being, the being which cares and is concerned for *itself*, because it is capable of misery and happiness.[22]

At present, it is crucial to see that for Descartes the *cogito*, because it follows universal doubt, offers a specific kind of information about this sort of thing that 'I' am, an information that cannot be gained otherwise. This knowledge is not only that *I am a thinking thing*. It is rather about how I (my 'self') stand in regards to this quality of being a thinking thing. As I am still actively denying the existence of everything else, I can see now clearly and distinctly that thinking cannot be something that is attributed to my being on top of other attributes or properties. All other possible attributes have been wiped out. So a thinking thing is, at this stage, *precisely, or uniquely* what I am: *Sum igitur praecise tantum res cogitans* 'to speak accurately, I am not more than a thing which thinks.' And this exclusive identification should also be true in temporal terms: *Quandiu autem?* 'But for how long?' *Nempe quandiu cogito*: 'For as long as I think.' Thus, my existence as 'self' lasts only *as long*, and *insofar*, as I think.[23] Contrary to what careless readers have often assumed, Descartes is not saying at this stage that I *have* no body, nor that I *am* not, actually, an extended being, besides being a thinking one. At the stage of the second Meditation, and Descartes will be adamant on that point in his Answers to the Objections,[24] these stronger assertions cannot be proved or disproved. A complete understanding of the real distinction of mind and body is not yet in order and will require a clear and distinct idea of the bodily substance. However, what already appears in the second Meditation is that I do not need to suppose that I have a body, or that I am an extended being, in order to see that I am something, a *subsisting*, substantial thing. Actual thinking is enough. Descartes is here making both an epistemic and an ontological point. The epistemic point is captured in the doctrine of 'complete things' which he will be at pains to explain to Arnauld and Gibieuf.[25] In the second Meditation, thanks to universal doubt, the mind is revealed as a *complete thing*, that is, it appears as 'a substance endowed with those forms or attributes which are sufficient for recognizing it as a substance'.[26] This form in that case is actual 'thinking'. The mere

[21] Descartes counts pain, pleasure, hunger, and in general sensations, as thoughts; but contrary to Locke, *pain* cannot be privileged in the *cogito* argument, as it is a confused idea that represents as existing something (a part of my own body) that may not actually exist (see the argument on the amputee, in P 4.196, AT 9, p. 315).

[22] In the first edition of the *Essay*, the term 'man' is used in this rather broad acceptation (man is both a body-and-soul *and* a person—see, for example, E I. iii. 5. The narrower definitions of *man* and *person*, contrasting their *principium individuationis*, only appear in the chapter on personal identity of the second edition (E II. xxvii. 21). There, the link between pain ('misery') and self-concern will become a prominent and specific aspect of the Lockean treatment of selves as *persons*.

[23] AT 7, p. 27: CSM 2, p. 18 (translation revised).

[24] See, for example, *Quintae Responsiones*, AT 7, p. 357: CSM 2, p. 247.

[25] See *Quartae Responsiones*, AT 7, pp. 219–28; CSM 2, pp. 154–60 and Descartes to Gibieuf (19 January 1642) AT 3, pp. 474–8: CSMK, pp. 201–4.

[26] *Quartae Responsiones*, AT 7, p. 222: CSM 2, p. 156 (translation revised).

thought that I am 'thinking' is then *sufficient* to make me distinctly recognize a substantial existence. Nothing else has been presupposed. By contrast, for example, a shape (or a movement) is not distinctly perceived as a complete thing, unless we see it embedded in the attribute of extension that it necessarily *presupposes* (even though we do not think always of extension, when we think—somewhat confusedly—of shapes or movements). The ontological point is the direct consequence of the epistemic one: as thinking is sufficient to make us see the mind as a substance, and as thinking does not presuppose any other attribute in order to be distinctly perceived, it can be said that *thinking* is the *principal attribute* of the thinking substance.[27] In other terms, it is its *essence*. As Marleen Rozemond writes, in a neat formula that encapsulates both points: 'the substantive contribution that the second meditation makes is then not just the (clear and distinct) idea that mind is a thinking substance, but the idea that it is sufficient for it being a substance that it thinks'.[28]

It seems to me that this very idea is foreign to Locke. Locke certainly would have sided along with Gassendi or Leibniz, when they suggested that there was a sophism or fallacy of some sort in the whole argument of the first two Meditations—and that it is simply not possible that hyperbolic doubt (an utterly artificial procedure) could substantiate such a strong ontological assertion as the one stating that thinking cannot be a mode of an extended substance.[29] At any rate, Locke's agnosticism on the nature of substances is not moved by Descartes's argument: we do not, and cannot by any means, know *what is sufficient for making a substance what it is*. Substance is a word for an idea which in itself, and by its own nature confuses all boundaries between things.[30] Awareness of oneself, however obtained, cannot help change this state of affairs.

So, in the Lockean version of the *cogito*, the knowledge of our own existence is a purely existential truth. It puts our *selves* into the realm of things, but that is all it accomplishes. Consciousness does not show us what sort of thing we are. It leaves us with the most ordinary idea of ourselves.

It is noteworthy that in the *Essay*, at least in the first edition, the ordinary reference for words such as 'I' or 'us' or 'our selves' is almost never the soul nor the 'mental thinking substance', but rather the whole man that we are and that we can recognize in a mirror, with its bodily envelope and whatever is enclosed in it, a digesting thing, a moving thing, a thinking thing. Locke indeed usually says that the thinking thing (the soul) is 'in us' rather than 'us': for example II. i. 10: 'there is something in us that has a power to think', IV. iii. 17 'the operations of our own souls within us' or IV. x. 19: 'that thinking thing within you'. Even in chapter II. xxvii he does not entirely renounce this

[27] See, for the definition of the principal attribute, *Principiae philosophiae*, I, 53 (AT 8, p. 25: CSM 1, p. 210).

[28] Rozemond, 'Descartes's Case for Dualism', p. 40, note 35.

[29] See Leibniz, *Critical thoughts on the general part of the principles of Descartes*, and Gassendi, *Objectiones Quintae*, in AT 7, pp. 257–8: CSM 2, p. 180. In chapter IV. iii. 6, Locke suggests that nothing in our idea of matter forbids that a material thing, fitly disposed, could think.

[30] See E II. xiii. 18. Because we do not have any determinate idea of what a substance is, we are led to a 'promiscuous use' of the term, applying it indifferently to God, Spirits, and Matter without being able to see whether it stands for the same idea or not.

way of talking, as can be seen in this striking sentence, where the reflexive us/ourselves stands both for the whole and for the part: 'in this ignorance we are in of the nature of that thinking thing that is in us, and which we look on as ourselves' (II. xxvii. 27).

So when Locke says in chapter IV. ix that 'we' know that 'we' are, I take it to mean in a rather prosaic sense: we as 'men',[31] rather than 'we' as *specifically thinking things*, are proved to exist. This can also be seen in this reformulation of the argument in the next chapter: '*Man has a clear perception of his own Being;* He knows certainly that he exists, and that he is something' (E IV. x. 1). Locke seems reluctant to equate this ordinary, unreflective, and almost trivial knowledge with the philosophical conclusion that there is a soul, a mental substance, *in us* which thinks. First-person conscious experience is certainly the *basis* for this later conclusion, but the inference of a mental substance comes only as a second step, that ordinary men—who are all certainly conscious of their own existence—may not take. Interestingly Locke does not mention the conclusion in chapter IV. ix which is only concerned with the 'plain' perception of our 'own existence'. The 'philosophical' idea of a spiritual being endowed with mental faculties appears only later, in chapter x, as something that we need for the demonstration of God's existence: 'Again a Man finds in himself *Perception* and *Knowledge*. We have then got one step farther; and we are certain now, that there is not only some Being, but some knowing, intelligent being in the World' (IV. x. 5).

Now there are certainly systematic reasons why Locke was reluctant to equate the knowledge we have of our selves through conscious experience, and the knowledge that there is a thinking substance in us which thinks. These cognitive operations are clearly distinguished—they differ both in their psychology, in their epistemic value, and also, I would suggest, in their object.

a. We mentioned already the psychological discrepancy between consciousness and reflection. They are not oriented the same way. Consciousness is a component part of any act of the mind when it is directed towards its object, its ideas—it is not a receptive faculty, it does not produce any idea. Reflection results from a specific act of the mind, a 'turning inward upon itself'. It is an internal sense, collecting the ideas of our own mental operations. As consideration of children shows, one may very well *think*, and thus *be conscious* of oneself, without considering, and perhaps not even having (in a dispositional sense) the (reflective) idea of ourselves as a *thinking being*, or a thinking substance, endowed with specific faculties (see E II. i. 8).

b. Conscious awareness and reflection also differ in their epistemic quality. The immediacy and intuitive certainty of self-awareness described in IV. ix contrasts with the rather obscure and convoluted inference concluding that a thinking thing exists in us. This inference, presented in the chapter on our complex ideas of substances (II. xxiii), is a complex, multi-step process. First, just as sensation passively receives ideas of external things, reflection passively receives the ideas of 'those operations of our own minds, which we experiment daily in ourselves, as thinking, understanding,

[31] See note 22.

willing, knowing, and power of beginning motion, &c' (II. xxiii. 15). Second, we have to realize that these 'faculties' or qualities of the mind cannot subsist by themselves, without something to support them (see LW 4, p. 29). This leads us to suppose in the third place the existence of some substratum, of which we know nothing except that it should be able to support and somehow causally produce these operative faculties. The inference is of the same sort as the one which leads us to suppose an unknown substratum for all sensible qualities and powers perceived as going 'constantly together' in the external world. In both cases, sensation and reflection, matter and spirit, the inference, far from being a self-evident truth, is a rather obscure supposition. The substratum is not perceived, nor intimated to our minds, it is only 'supposed' or 'postulated'. It is something that we have to posit in order to give coherence to the way our perceptions (whether external or internal) are structured, but, as such its essence is entirely unknown to us.

c. Finally a case can be made that self and thinking substance are not identical. The consciousness aspect of my thoughts proves directly that I (*my self*) am; reflection on my mental operations forces me—rather indirectly—to recognize that there is something, somewhere in the world (and most probably, down here, inside my body), which thinks. This leaves quite open the issue whether it is right to identify these objects. Am I (my 'self') identical with the mental substance that I am led to suppose exists behind my very faculties? The Cartesian answer of the *Discourse on Method* and *Meditations* is: I am *precisely* that, a mental, thinking substance. But Locke's own construal of the whole issue cannot deliver such a direct answer. The prosaic account of self that is dominant in the first edition (self is this full carnal package, with its occasional thoughts, its absentmindedness, its pleasures, and its pains), allows to say only this: there is something *in us* that thinks, this thing belongs to us for sure, but it is not clear how much we need it, and whether it is sufficient for being 'my self'. Chapter II. xxvii of the second edition will precisely delve into this uncertainty.

3. Thinking not the Essence of the Soul

In a digression of chapter II. i, one of the most explicitly anti-Cartesian passage of the *Essay*, Locke takes issue with a seemingly absurd consequence of the second Meditation, namely that it should be factually true, if actual thinking is indeed essential to the soul, that the soul always thinks.[32]

Descartes answers a similar objection of Gassendi's in the Fifth Replies (AT 7, p. 356), as well as in his 1641 Letter to Hyperaspistes (AT 3, pp. 423–4). He denies that there is any absurdity in the consequence, and simply accepts it: 'You say you want to

[32] Descartes acknowledges this consequence in a Letter to Arnauld, 4 June 1648, AT 5, p. 193: CSMK, p. 355: 'Nevertheless it seems necessary that the mind should always be actually engaged in thinking; because thought constitutes its essence, just as extension constitutes the essence of a body. Thought is not conceived as an attribute which can be present or absent like the division of parts, or motion.' See also La Forge, *Treatise on the Human Mind*, ch. 6, pp. 56–61.

stop and ask whether I think the soul always thinks. But why should it not always think since it is a thinking substance?' (AT 7, p. 356; CSM 2, pp. 246–7).

Locke's argument is well known (if not always easy to understand)[33] and I shall not rehearse it here in all its details. It seems to follow two partly contradictory strategies. On the one hand, the argument is directed against Descartes's supposed dogmatism: the question of whether my soul always thinks being a purely factual, empirical one and also a *private* one, it cannot be decided a priori, on the basis of a dogmatic definition of the soul, made by someone else. This, at least, should be true for anyone who, like Locke *and* Descartes, accepts the idea that consciousness is inseparable from thought, and that first-person (conscious) experience is the only gauge for judging whether something is, or is not, a thought. Souls may very well not do any thinking, for example during sound sleep—Locke humorously suspects that this might be the case for his own 'dull' soul. But, and here we have the thrust of the argument, as no one can experience his own soul when it is not thinking, the question appears eventually to be unanswerable through inner first-person experience, and so we should be content here with strict agnosticism.

However, there is a second line of argument, in which Locke seems less prone to agnosticism, and where he strongly suggests that the Cartesian view is not only dogmatic, but very likely false, in that it leads to a representation of thought, moral persons, and human beings, which is unpalatable. The argument here is quite clearly a 'doxastic' one: even if we cannot *demonstrate* that the soul thinks intermittently rather than always, the former opinion appears to be more attuned to our moral, religious, and anthropological needs, and so it should be *preferred*. This perspective occupies the most part of the long digression of II. i and is argued by means of various puzzling cases and examples: Socrates asleep, Socrates awake (II. i. 11); Castor and Pollux sharing the same soul, awake and asleep in turns (II. i. 12). These cases test the possible consequences of the Cartesian positions and justifications, and somehow stretch them to the limit. Locke here clearly argues *ad hominem*, availing himself of Cartesian (or pseudo-Cartesian) premises, such as the real distinction of the soul and body, the aptitude of the soul to think 'apart' from the body, the doctrine of beast-machines, etc.[34] As we have no memory of having been always thinking, we are led to suppose (if we are to defend the Cartesian thesis) that actual conscious thinking sometimes occurs in the soul, without leaving any trace of itself, and so without having any consequence for the embodied man whose soul thinks in this mode.[35] The soul thinking on its own (for example, in sound sleep) might have 'memories' of its own, desires and volition of its own, which the man who is asleep is never conscious of.[36] In that case,

[33] For an illuminating discussion, see Jolley, 'Dull Souls and Beasts'.

[34] Cf. E II. i. 12: 'Let us suppose, then, the soul of Castor, while he is sleeping, retired from his body; which is no impossible supposition for the men I have here to do with, who so liberally allow life, without a thinking soul, to all other animals.'

[35] E. II. i. 15: 'to think often, and never to retain it so much as one moment, is a very useless sort of thinking.'

[36] Cf. E II. i. 12 and II. i. 19.

it would be appropriate to say that there are two people housed in one man. Locke does not push this point very far. He is simply satisfied in presenting this strange hypothesis to his reader, strongly suggesting that there is something morally unacceptable in such a representation. If we weren't the same person as our soul (the being that supposedly persists *post mortem* and is judged) it would be impossible to rightfully assign merits and demerits. We might be pardoned or condemned for thoughts and volitions which we were never properly conscious of.

The whole discussion opens with the striking statement according to which 'the perception of ideas [is] (as I conceive) to the Soul, what motion is to the Body, not its Essence but one of its Operations' (II. i. 10). This rather bold metaphysical assertion seems to fly in the face of Locke's official agnosticism on metaphysical matters,[37] but it is clear that its intent is not to give any positive account of the essential nature of the soul, but rather to criticize what Locke takes as a typical expression of Cartesian dogmatism: the view according to which, thinking being the essence or 'principal attribute' of the soul, whatever pertains to the soul has to be a mode of thinking. To Locke, assuredly, this is too strong an ontological constraint: a thinking thing may have, for all we know, properties that are not presupposing the possession of thought. It may be the case that actual thinking is something that souls do *usually* or, even perhaps the *only thing* they do, or a thing that *only they* do,[38] but nevertheless still not something they do necessarily (or always[39]). The point is illustrated in the analogy of motion. A body can move, or it can stay at rest. There are some bodies that do not move, and bodies that move, are not always moving. Motion, actual motion, is not essential, because it is not always presupposed in our representation of body.[40] I can have clear and distinct idea of a body without supposing that it is moving (although I cannot have it without supposing that it is either moving or at rest).[41] The point is the same for thinking. It is not conceptually incoherent to suppose that 'non thinking' episodes take place in a 'resting' thinking substance. The power or faculty of thinking can be exerted or not; thoughts are operations and as such their existence in the thinking thing is contingent. Of course, we cannot observe in our minds, and even less in other minds, episodes of pure rest, of pure absentmindedness—the mind is only observable

[37] See the discussions in Bennett, 'Locke's Philosophy of Mind', and Jolley, 'Dull Souls and Beasts'.

[38] In those cases, actual thinking would be a *proper* attribute of the soul, in one or the other of the various senses of the term in the Porphyrian theory of predicables. Locke significantly resorts to this terminology in the next sentence, when he suggests that thinking may well be 'the proper action of the soul'.

[39] In a note to his French translation (EEH II. i. 15, p. 69), Coste remarks that Locke is perhaps overstating his point: even if thinking is not the essence of the soul, it might be factually true that it always thinks. Coste mentions in his note that Samuel Clarke made this point in *A Defense of Dr Clarke's Demonstration of the Being and Attributes of God* (London, 1732).

[40] This at least is true at the level of ordinary physical experience, and it seems that it is here all that Locke wants to consider. Leibniz expresses his disagreement in the *New Essays*, making a point that actual motion (however infinitesimal) should be an intrinsic features of bodies. See Leibniz NE, A 6, 6, p. 112: '*le corps n'est jamais en repos*'.

[41] In this sense, motion is not, strictly speaking, a primary (or inseparable) quality. Locke sometimes refers to it as such, but what he has in mind is always the somewhat 'gerrymandered quality' *motion or rest*. See Jacovides, 'Locke on Primary and Secondary Qualities', p. 102.

when it thinks. For that reason, thinking and not thinking are not exactly on the same 'epistemic' foot as motion and rest. In a sense, we don't know that thought is only contingent or accidental to the thinking substance. Nevertheless, and quite independently of the doxastic results of the puzzling cases, the very experience of thinking (at least in Locke's descriptions) strongly supports the thesis that it is an action rather than the essence of the mind.

The point is clearly made in chapter II. xix. 3 where Locke remarks that our thinking is liable to various degrees of attention, ranging from the higher degree of 'intention' and 'earnestness', to an almost complete passivity, a 'very near minding nothing at all'. This consideration brings Locke back to the question of the essence of the soul:

> This difference of *Intention*, and *Remission* of the mind in thinking, with a great variety of Degrees, between earnest Study, and very near minding nothing at all, Every one, I think, has experimented in himself. Trace it a little farther, and you find the mind in Sleep retired as it were from the Senses, and out of the reach of those Motions made on the Organs of Sense, which at other times produce very vivid and sensible ideas.
>
> ...Since, I say, this is evidently so in matter of Fact and constant Experience, I ask whether it be not probable, that *thinking is the Action and not the Essence of the soul*? Since the Operations of Agents will easily admit of intention and remission; but the Essences of things are not conceived capable of any such variation. (E II. xix. 4)

Quite remarkably the argument is couched in terms that are strongly reminiscent the old scholastic doctrine of the *intentio and remissio formarum*, and this is a quite rare example of Locke using scholastic concepts to make a metaphysical point. The doctrine of 'latitude of forms' may be traced back to Buridan, Oresme, and Henri of Gand. It was used precisely in order to distinguish between essential qualities (or forms)—such as 'humanity', which does not admit of any intention or remission, and accidental ones, such as heat, which do.[42] Locke suggests here that thinking is just like heat: it is an accident, because one could have more or less of it, according to the degree or intensity of the attention that we direct on what passes in our minds. This is a remarkable (if not entirely convincing[43]) argument. It partially answers the main difficulty of the intermittency thesis: gaps in consciousness are no objects for direct experience. However, as we experience variations in the intensity of our conscious awareness, we could see how psychologically plausible it is that sound sleep coincides with a complete remission of thinking. This, in any case, is the most direct argument (and perhaps the only one that is not constructed *ad hominem*) against the Cartesian thesis.

Let us conclude here with chapter II. i. The long discussion of thinking souls was not placed by chance at the very beginning of Locke's enquiry into ideas and knowledge. Even more clearly than Book I on innate ideas, it was intended as an anti-Cartesian

[42] For an interesting discussion of its use in Locke in contrast with Descartes's mention of it in the opening pages of the *Discourse on Method*, see Simonetta, 'L'Âme pense-t-elle toujours?'

[43] Leibniz, who is also applying a continuity principle to his analysis of thought, would consider that the argument pleads for unconscious thinking rather than for intermittency.

manifesto, with an important role to play in what Jolley has rightly described as a *systematic* anti-Cartesian metaphysical programme.[44] The attack against the apparently secondary Cartesian thesis 'the soul always thinks' is an indirect strategy intended to convey to the reader a determined metaphysical stance, which concerns the *Essay* as a whole. The soul is not essentially but only accidentally a thinking being. Locke identifies in Descartes's essentialism the basis for a metaphysical delusion, namely that of an entirely independent being that could exist apart from the body and could think somehow unconditionally in a purely intellectual manner. The whole Lockean outlook, epistemological as well as anthropological, may be described as a reaction to this view of a disembodied mind. The most abstract and most remote among our ideas has still its origin in the senses, which means that it owes something to the body. The exercise of thinking is limited in scope and penetration because the organs of senses that provide its ideas are themselves limited in number and acuity; specific trains of ideas certainly depend on the structure of the brain, and on how it allows the stream of 'animal spirits' to pass here or there; it is affected by pathological associations (II. xxxiii), subject to remission (in episodes of absentmindedness), to interruption in sleep, and to termination in death. It should be recognized that corporeity is also, positively, what permits us to make something of our thinking. Thoughts which do not leave in a brain any material traces of themselves, whether they take place in an infant, a sleeping soul, or in a separated mind, are no nobler than the chaotic motion of atoms in an Epicurean world, or 'characters drawn on dust that the first breath of wind effaces' (II. i. 15). They are a vain rhapsody of disconnected perceptions. As they cannot be retained, they cannot be put to use, converted into memory, experience, or knowledge. In brief, we are finite corporeal and dependent Creatures. This state affects not only our life, as corporeal mortal beings, but also as intellectual ones; we should not overestimate our thinking, but should recognize its limitations, if we wish it to be rational and reasonable.

One should perhaps add that the attacks against Descartes's concept of the mind are not immune from misrepresentation or caricature. It was certainly not in Descartes's intention to deny our finitude as thinking beings, and he never properly said that the human mind, in the condition which is ours here below, can entirely dispense with bodily affections in order to reason and develop intellectual thinking.[45] Instead, his point is that some of our ideas (such as the thought we have of ourselves) are originally in us, and were never abstracted from matter nor copied from bodily representations. Access to such ideas requires *silencing* the suggestions of the senses and of bodily representations: a task which often requires an intense effort ('contention') of the mind applied *against the body*.

Locke's puzzling cases offer other examples of misrepresentation. They heavily draw on the idea that Cartesian human souls and Cartesian human bodies are separable in principle and could live and think separately. But neither Descartes's 'real distinction'

[44] See Jolley, 'Lockean Abstractionism'.

[45] See Kambouchner, 'Descartes et l'indépendance de l'esprit', and Kambouchner, *Descartes n'a pas dit*, pp. 81–9.

nor the animal-machine suggestion were primarily aimed at demonstrating the separability of the soul in human beings. First and foremost, Descartes was attacking the scholastic Aristotelian theory of knowledge, which held that the human immaterial soul (even though it may be separated from the body through a divine act) cannot think, in this terrestrial state, without abstracting its thoughts from bodily impressions and images.[46] Of course, what is really distinct can be separated by God's power, but an excessive stress on separability leads one to conceive the 'union' of mind and bodies in men as somewhat accidental—like the pseudo-union of ship and pilot. Descartes knew that his mind–body doctrine could easily be interpreted that way, but he forcefully rejected this interpretation of the union.[47] Although soul and body are really distinct, they are made in such a way that one is, so to speak 'instituted' for the other: the body is available for the soul in voluntary motion, and the soul for the body in the affections of the senses and the passions. 'Insofar as the man is considered in himself as a whole, we say that he is entirely *ens per se*, and not *per accident*, because the union by which mind and body are joined to each other, is not accidental to him, but essential, since without it a man is not a man.'[48]

These considerations show that Descartes's subtle doctrines were often quite far away from the *doxa* of substance dualism (or separatism) that Locke is attacking. Indeed, although Descartes is explicitly targeted, the doctrine under attack resembles more a Christian Platonic doctrine, of the sort defended at the time by someone like Henry More, insisting on the autonomous life (psychozoia) and immortality (psychathanasia) of the soul.[49] I suggest that we keep this important proviso in mind in examining Locke's treatment of the identity of 'persons' in chapter II. xxvii.

4. Persisting Selves

In the vast literature on chapter II. xxvii, it has been often assumed that Locke specifically targeted Descartes's view of personal identity—or rather, as Descartes never really discusses the issue, a somewhat reconstructed view of what it takes to be a Cartesian person. For example, Henry E. Allison argues that 'It was nothing other than his fundamental opposition to the dominant philosophy of Descartes which posed the problem for Locke. For Descartes the indivisibility of the self or thinking substance is self evident truth and from this rationalistic standpoint the problem of personal identity never really arose.'[50]

46 The point is persuasively made in Rozemond, *Descartes's Dualism*, ch. 2.

47 See especially the *Quartae responsiones*, and the letters to Regius of December1641 and January 1642.

48 Descartes to Regius, January 1642, AT 3, p. 508: CSMK, 209.

49 McCann, 'Cartesian Selves and Lockean Substances' similarly argues that Locke's anti-Cartesian arguments hold against a view of immateriality that is closer to Henry More's doctrine of separate or separable souls, than to Descartes's actual construal of immateriality.

50 Allison, 'Locke's Theory of Personal Identity', p. 41.

I think Allison is right when he suggests that Descartes was not attuned to the problem of personal identity. More intriguing perhaps is the fact that he situates the reason for that in the self-evident character of the indivisibility of the self. I take Allison's implicit argument to be something like this: if I have an absolute certainty that the thinking substance which in different times I recognize as my self is always the same thinking substance, on account of its essential *indivisibility*, then I have secured a metaphysical criterion for the identity of the self, and I do not have to raise the issues of consciousness, memory, accountability, in order to understand what makes a person the same person. By contrast, Locke does not consider anymore that we have a self-evident or clear and distinct idea of the nature of a thinking substance. So he cannot entirely rule out the possibility that our soul is a compound substance, divided into successive parts (as it would be if we imagine that various thinking substances take their turn in the same man). We don't even know with certainty whether our soul is an immaterial substance. According to the suggestion of IV. iii. 6, it could be a material, corruptible, and divisible one, to which the faculty of thinking had been 'superadded.' This agnostic stance concerning the nature of the soul awakens us to the problem of personal identity: there is no guarantee that our identity is necessarily attached to the persistence through time of one individual immaterial soul. And this reveals clearly that our concept of personal identity, if it has any consistency, cannot be based on the consideration of the persisting through time of 'the same' thinking substance. Rather, as Locke intends to show, it should be grounded on the fact that our thoughts now and then all partake in the same 'continued consciousness'.

Before trying to test the worthiness of such a reconstruction, I would like to draw attention to the fact that in the chapter on Identity, Descartes is not as conspicuously present as one would expect. Of course, there are some signs that these pages are still in line with the anti-Cartesian outlook of chapter II. i. Perhaps the strongest is the fact that Locke's argument against the *same-immaterial-substance* view of identity resorts to almost identical puzzle cases, as the one used for demonstrating that the soul does not think always, considering detached souls going back and forth in various independent bodies. The cases are similar, but the use of them, it should be noted, is significantly different—they are not any more intended as illustrations of the potentially absurd consequences that one would have to swallow if one wants to defend the Cartesian doctrines. Instead, they are thought-experiments used to drive home a conceptual point about how we should understand the notion of personal identity, and whose absurdity or impropriety is not considered anymore as pertinent.[51] Descartes's specific view on the essence of the soul or his doctrine of beast-machines are not mentioned, nor even alluded to in chapter II. xxvii, which, however, does mention 'Christian Platonist' and 'Pythagorician' views of detached souls and metempsychosis. So if, as Allison says,

[51] Cf. II. xxvii. 27: 'I am apt enough to think I have, in treating of this Subject, made some Suppositions that will look strange to some Readers, and possibly they are so in themselves. But yet I think, they are such as are pardonable, in this ignorance we are in of the Nature of that thinking thing, that is in us, and which we look on as our *selves*.'

fundamental opposition to Descartes is the driving force here, it is a question why Locke retreated from a direct confrontation with the views of his predecessor.

I think we can answer this in considering the context in which what Allison calls the 'problem' of personal identity first appeared to Locke. This context is, most probably, a theological one: the discussion of immortality. In an important note from his journals, dated from 20 February 1682, Locke considers and criticizes what he calls 'the usual, physical proof of immortality' in terms that have a direct bearing with the whole discussion of II.xxvii. His point is that it is especially important, for religious and moral reasons, that an account of *immortality* should explain not only the subsistence of the soul but the persistence of a state of consciousness and sensibility, so that the resurrected man would be punished or rewarded for what he did, and for what he is now conscious of having done. Locke suggests that 'the usual proof' precisely failed in this respect. He describes it thus:[52]

> The usuall physicall proofe (as I may soe call it) of the immortality of the soule is this, Matter cannot thinke ergo the soule is immateriall, noe thing can naturally destroy an immateriall thing ergo the soule is naturally immortall.

But, as Locke explains then, it is not enough nor perhaps necessary to prove that the soul is immaterial in order to prove that it is (in a morally relevant sense) immortal.[53] The kind of subsistence that is 'usually' proved is, according to Locke's terms, 'bare substantial subsistence', and as such it is not worthier of our consideration than the bare incorruptibility and subsistence of Epicurean atoms. This sort of subsistence is contrasted with the persistence of what Locke calls then *a state of sensibility*, or a state of life and perception, which is alone capable of ensuring the consciousness of past thoughts and acts.

Interestingly, the argument that bare substantial existence is not sufficient to ensure the moral requisites of immortality is based on the conviction that the thinking substance does not always think:

> Since, then, experience of what we find daily in sleep, and very frequently in swooning and apoplexy, etc., puts it past doubt that the soul may subsist in a state of insensibility, without partaking in the least degree of happiness, misery or any perception whatsoever (and whether death, which the scripture calls sleep, may not put the souls of some men at least into such a condition, I leave those who have well considered the story of Lazarus to conjecture), to establish the existence of the soul will not prove its being in a state of happiness and misery, since it is evident that perception is no more necessary to its being than motion is to the being of body.

[52] MS Locke f. 6, pp. 25–6, quoted in AG p. 128.

[53] See E IV. iii. 6: 'All the great Ends of Morality and Religion, are well enough secured, without philosophical Proofs of the Soul's immateriality; since it is evident, that he who made us at first begin to subsist here, sensible intelligent Beings, and for several years continued us in such a state, can and will restore us to the like state of Sensibility in another World, and make us capable there to receive the Retribution he has designed to Men, according to their doings in this Life.'

This passage provides the key connection between chapter II. i and II. xxvii, and it also explains why Descartes's doctrine of thinking substances *could not* be directly addressed in chapter II. xxvii. It is clear now that it is only because Locke allows thinking to be only an accident of the thinking substance that the soul may very well subsist 'substantially' after death, but in a state of complete insensibility. In fact, as Locke suggests here, even during this life, we can be said to experience many deaths and resurrections: when the soul ceases to think for a while (for example in sound sleep) and then awakens and picks up the thread of its former consciousness, it is in a sense just as if we have been raised from the dead, with the renewed consciousness of our past actions. Thus, the analogy from daily experience is put forward as evidence that God has the power to renew the consciousness of our past actions at any time. If He can do so in our (changing) body in this life, He can also do so (in another body) in the next.

This will make it clear that the substantialist account of personal identity that Locke criticizes is grounded not so much on a Cartesian but on a Lockean view of the soul: the soul is only intermittently thinking, and this is why it cannot ensure personal identity. If Descartes were right in his account of souls, there would be no problem at all with the substantialist account. The thinking substance would be by its very nature always in a state of sensibility, always conscious of itself and, consequently, it would ensure personal identity, just as consciousness does according to Locke.

Locke certainly realized that he could not have it both ways: if Descartes deserves criticisms for holding that the thinking thing is always conscious, he cannot be criticized for omitting consciousness in his account of persisting souls. And this explains why Descartes, who certainly was still in Locke's mind when he wrote the chapter, does not appear any more, at least textually, in II. xxvii.

But now we come close to a tempting suggestion: although Locke would never dare to admit it, it seems that, together with the new understanding of persons in chapter 27, something of a genuinely Cartesian spirit has surreptitiously entered the picture of Lockean philosophy of mind.

Let us recall the most characteristic thesis of Locke's account of persons: what makes a person (or a self[54]) the same person at different times, is neither the persistence through time of the same body nor the persistence of the same immaterial soul, but the fact that we are now and then 'agreeing in the same consciousness'. The exact meaning of this is not straightforward. I contended elsewhere[55] that the consciousness criterion is not, contrary to a widespread opinion, a memory criterion. If I am now the same person that I was then, it is not because I now remember my past thought, but rather because my present and past thoughts belong both to the same unique and continuous train of

[54] The terms are, in most cases, interchangeable in Locke's usage of chapter II. xxvii, as can be seen in this set of definitions: 'we must consider what person stands for;—which, I think, is a thinking intelligent being, that has reason and reflection, and can consider itself as itself, the same thinking thing, in different times and places' (§9). 'Self is that conscious thinking thing…' (§17). 'Person, as I take it, is the name for this self' (§26).

[55] Hamou, 'Mémoire et "Conscience continuée"'.

conscious ideas, the same 'continued consciousness'. Memory has indeed a unique privilege in that it alone makes me aware of my personal identity through time, but it does not constitute it, as one can see from the fact that we never actually remember all the thoughts we consciously had, and also that we sometimes forget them, or misrepresent them.[56] So Locke's grand thesis, as I understand it, is based not on memory but on the psychological fact of the *train of ideas*: my soul may not always think, thinking may be interrupted, and renewed periodically, but still my thoughts will always be *consciously* represented to me as belonging to one single string of actual successive ideas, always going on, and whose duration measures exactly the duration of my own existence as conscious being. This temporalized *cogito* is expressed in a striking statement of II. xiv. 3—a passage surprisingly rarely commented on in the literature on personal identity:

> For whilst we are thinking, or whilst we receive successively several *Ideas* in our Minds, we know that we do exist; and so we call the Existence, or the Continuation of the Existence of our selves, or anything else, Commensurate to the succession of any *Ideas* in our Minds, the *Duration* of our selves, or any such other thing co-existent with our Thinking.

What kind of ontological consistency shall we impart to this continual train of ideas (later described as the 'stream of consciousness'), commensurate to the existence of ourselves? Clearly for Locke not that of a substance: ideas are accidents or modifications of the mind, they do not constitute it. In themselves they are transitory and clearly dependent beings, just like modes are,[57] in need of support or substratum either spiritual or corporeal. Locke never changed his mind on that. Still, one may wonder whether his account of the identity of person through the notion of a continued consciousness does not require considering an entity that is somewhat more than a mode, if less than a substance: a lasting, indivisible, and continuous being of which our actual ideas are no accidents but rather essential, constitutive, parts. This being is what Locke calls 'consciousness' in chapter xxvii—although not in the sense already mentioned of an individual act of self-awareness accompanying each one of our thoughts, but rather

[56] In II. xxvii. 10 indeed, the limited scope and fragility of memory is presented as one of the main reasons why we need a theory of personal identity that goes beyond a theory of substantial identity.

[57] Modes 'contain not in them the supposition of subsisting by themselves, but are considered as dependences on, or affections of substances' (E. II. xii. 4). LoLordo's *Locke's Moral Man* makes a good case for considering persons as *modes* rather than substances, thus reviving the old interpretation put forward by Edmund Law in the eighteenth century. According to LoLordo, the fact that persons have subjective authority on their own constitution, and the fact that Locke discusses the nature of persons through conceptual analysis (puzzle cases), rather than through the observational, descriptive method, shows that person is, like mixed modes, something we impose on the natural world to serve our own purposes, rather than something we copy from nature. LoLordo concedes that Locke often says of persons that they are 'beings', or 'thinking being', and this seems to plead for the substantialist account, but to LoLordo persons are substances only through metonymy, in the same indirect way as a 'father' is a substance: what makes me a father/person is a modal relation *superinduced* to the human substance that I am.

as the concatenation of these successive conscious thoughts in one single continuous train of thought.

So here is the suggestion: it is difficult not to be struck by strong similarities between the specific account of continued consciousness that we find in II. xxvii and crucial features of the Cartesian *thinking thing*. Let us here just list a number of basic characteristics which they both share. They are both identified as an intellectual being, which exist insofar and as long as it thinks. Locke takes issue with Descartes on the thesis that the *soul* always thinks, but he himself acknowledges that it is in the nature of *consciousness* to allow no gaps in the train of ideas. The 'soul' may be intermittently thinking, with temporary periods of rest, or sound sleep, but consciousness always appears to itself as one continuous whole, the last conscious instant before the soul goes to sleep being always connected with the first conscious instant of awakening.[58] Both are (in this life) constantly, and strongly attached or united to an animated body, that they have all reasons to consider as *theirs*, but nevertheless both are *conceptually independent* from it, because the persisting conditions of the thinking thing can be *conceived* independently of the persisting conditions of the living body. In order to establish this, both Descartes and Locke resort to hyperbolical thought experiments: puzzle cases and universal doubt play, on this score, a similar role. The body, and, for all we know, its 'soul', or whatever supports its intellectual 'faculties', might be changed, or even annihilated, *I* shall stay the same Cartesian thinking thing or the same Lockean person as long as I am aware of my successive actual thoughts.

This, in turn, reveals another striking similarity: subsistence through time is not something that is presupposed in our concept of the self. In fact, both Locke and Descartes tend to think that the continuous existence of the same self is entirely in the hands of God. We saw this clearly expressed in Locke's account of death and sound sleep. But it is no less explicit in Descartes who, for example, explains to Arnauld, that there is nothing in the idea of myself that forces me to think that I could not cease to exist the next second.[59] I take this to mean that the inherent *successiveness* of my ideas is something that is not under my control, any more than the passage of time, although it surely makes me the being that I am. What makes in Descartes created substances persist in time is not their 'bare substantial nature' but, for all we know, the very same act of Creation that allows them to exist in the first place.[60] In one place, Descartes

[58] E II. xiv. 4: 'When that succession of *Ideas* ceases, our perception of Duration ceases with it; which every one clearly experiments in himself, whilst he sleeps soundly, whether an hour, or a day, a month or a year; of which Duration of things, while he sleeps or thinks not, he has no perception at all, but it is quite lost to him; and the moment wherein he leaves off to think, till the moment he begins to think again, seems to him to have no distance.'

[59] 'We clearly understand that it is possible for me to exist at this very moment, when I am thinking of one thing, and yet not to exist at the very next moment, when if I do exist, I may think of something quite different' (Descartes for Arnauld, 4 June 1648, answering to the suggestion that the duration of a spiritual being is not successive, AT 5, p. 193: CSMK, pp. 354–5). See also Third Meditation, AT 7, p. 49.

[60] In what is usually called the third proof from effects (Third Meditation and *Principia* I, 21), Descartes argues that the very fact that I subsist in time is a proof of God's existence, since I do not find in myself the power to recreate myself in each and every instant of my life.

even said that he would be easier for him to conceive a thinking substance existing intermittently rather than a thinking substance existing without thinking.[61]

This point is important because it explains why Locke's criticisms somehow miss the mark in relation to the Cartesian 'substantiation' of the self. Descartes is no prisoner of any traditional view of substance. He never suggested that permanency or endurance are defining features of thinking substances. Similarly, he never considered immateriality as some sort of substantive property of the thinking substance that could be attributed to it independently of its actual thinking—an 'unsolidity' or 'in-extendedness' which would describe the 'bare substantial existence' of the immaterial stuff out of which thinking things are made. Such an account of immateriality, which I suppose Locke is tacitly endorsing, especially when he criticizes Descartes and the Christian Platonists, fails to offer any specific reason to consider why an immaterial soul is better qualified than a material one to receive the power of thinking.[62]

But obviously this was not how Descartes construed the substantiality of the soul. In fact, what makes a soul a substance is not so much its endurance and immateriality as its independency and indivisibility. The nature of substantiality is not predefined in terms of the persisting existence of an inert substratum, it is discovered in the very act of the *cogito* which reveals the self to itself as an *independent* thinking being.

In the synopsis to the Meditations, Descartes suggests an ontological contrast between bodies and minds in regards to their substantiality:

> we cannot understand a body except as being divisible, while by contrast, we cannot understand a mind except as being indivisible. For we cannot conceive of half of a mind, while we can always conceive of half of a body, however small, and this leads us to recognize that the natures of mind and body are not only different, but in some way opposite. (AT 7, p. 13: CSM 2, pp. 9–10)

Why could not the mind be a bundle of thoughts, as the body is a bundle of material parts? A few lines later, Descartes explains that the mind is 'a pure substance' precisely because it remains the same whatever the changes occurring in its accidents (its thoughts). Although my mind has several successive thoughts, in each one of these thoughts, I remain the same self, and this unity of consciousness is precisely what makes me know that I am a pure substance. What is striking here is the fact that it is not on

[61] Descartes to Gibieuf, January 1642, AT 3, p. 478: CSMK, 203: 'it would be easier for me to believe that the soul ceased to exist at the time when it is supposed to cease to think than to conceive that it existed without thought'.

[62] See the striking thought experiment that Locke submitted to Stillingfleet (LW 4, p. 464): 'God has created a substance: let it be, for example, a solid extended substance: is God bound to give it, besides being, a power of action? That I think, nobody will say. He therefore may leave it in a state of inactivity, and it will be nevertheless a substance; for action is not necessary for the being of any substance, that God does create. God has likewise created and made to exist, de novo, an immaterial substance, which will not lose its being of a substance, though God should bestow on it nothing more but this bare being, without giving it any activity at all. Here are now two distinct substances, the one material, the other immaterial, both in a state of perfect inactivity. Now I ask, what power God can give to one of these substances (supposing them to retain the same distinct natures, that they had as substances in their state of inactivity) which he cannot give to the other?'

account of its substantiality that the mind is considered as indivisible or unique, but rather because of its indivisibility (construed as unity or identity through time of the same consciousness) that it is conceived as substantial. The unity of bodies, by contrast, can be destroyed through division, because their individuation or identity principle is a certain configuration of material parts that can be easily altered through a modal action. A body, in this sense, may not be considered as a 'pure' substance, even though the whole *expansum* of matter, or any one of its parts, is.[63] It is important to note that the indivisibility of the mind features as the central point in Descartes's 'proof' of immortality sketched in the Synopsis. As the soul cannot be divided into parts, it cannot be destroyed in the sense a human body can: the human body perishes, not because it is material (matter as such is as incorruptible as spiritual substances) but rather because it is divisible and, in this, lacks substantiality.

So, we are left here with two very different readings of Locke's relationship to Descartes's conception of selves and thinking substances. In one, Locke argues against what may be described as the Cartesian 'substantiation' of the self: the self is an immaterial thinking substance, whose nature is to think always. But he does so in using what appears to be only a pseudo-Cartesian (and somehow more Lockean) concept of a substance—that is, a *something we don't know what* that we need to posit in order to ground the power of thinking, and to account for its *subsistence*. The second reading brings Descartes and Locke closer, if not wholly in terms, at least in insight, in the sense that in both authors, what makes our self *a same indivisible and continuous being* is just what makes our consciousness a same consciousness. Whether or not we want to use the language of substance in order to describe the temporal stretch of conscious thoughts that supports the existence of persons is perhaps immaterial. What counts is that some being is here identified. This being may *be united to* a body, or to whatever material or immaterial fluids or spirits we need to constitute the 'soul'. Locke and Descartes would agree that it is irrelevant to what it is. This second 'Cartesian' reading would make Locke a paradoxical—and no doubt involuntary—heir to one of the deepest and perhaps less understood aspects of Descartes's doctrine of mind.[64]

References

Alanen, Lili. 'Thought', in L. Nolan, ed., *The Cambridge Descartes Lexicon*. Cambridge: Cambridge University Press, 2015.

Allison, Henry, E. 'Locke's Theory of Personal Identity: A Re-examination', *Journal of the History of Ideas*, 27 (1966), 41–58.

[63] On the distinction between the non-substantiality of individual bodies (whose principle of unity is modal) and the substantiality of any individual parts of extension, see in this volume Martha Bolton's contribution, and Garber, *Descartes's Metaphysical Physics*.

[64] Primitive versions of this chapter were read in Lille, Kilpisjarvi, and Berlin. I would like to thank, for their perceptive comments, Vili Lähteenmäki and the warm audience of the 2013 arctic circle seminar; Dominik Perler and his early modern group at the Humboldt University, Martine Pécharman, John Milton, Sandrine Roux, as well as two anonymous Oxford readers.

Arnauld, Antoine. *Des vraies et des fausses idées* (1683), ed. D. Moreau. Paris: Vrin, 2011.

Ayers, Michael. 'The Idea of Substance and Power in Locke's Essay', *Philosophical Quarterly*, 25/98 (1975), 1–27.

Balibar, Étienne. 'Introduction', in John Locke, *Identité et différence: An Essay Concerning Human Understanding, II. xvii. L'invention de la conscience*, ed. E. Balibar. Paris: Le Seuil, 1998: 9–101. (English translation, *Identity and Difference: John Locke and the Invention of Consciousness*, London: Verso, 2013.)

Bennett, Jonathan. 'Locke's Philosophy of Mind', in V. C. Chappell, ed., *The Cambridge Companion to Locke*. Cambridge: Cambridge University Press, 1994: 89–114.

Chalmers, David J. *The Conscious Mind in Search of a Fundamental Theory*. Oxford: Oxford University Press, 1997.

Coventry, Angela, and Kriegel, Uriah. 'Locke on Consciousness', *History of Philosophy Quarterly*, 25/3 (2008), 221–42.

Garber, Daniel. *Descartes's Metaphysical Physics*. Chicago: University of Chicago Press, 1992.

Hamou, Philippe. 'Mémoire et "conscience continuée": une lecture de Locke sur l'identité personnelle', *Philosophical enquiries. Revue des philosophies anglophones*, 3 (2014), 1–33. <http://www.philosophicalenquiries.com>.

Jacovides, Michael. 'Locke on Primary and Secondary Qualities', in Lex Newman, ed., *The Cambridge Companion to Locke's Essay Concerning Human Understanding*. Cambridge: Cambridge University Press, 2007: 101–29.

Jolley, Nicholas. 'Lockean Abstractionism Versus Cartesian Nativism', in Paul Hoffman, David Owen, and Gideon Yaffe, eds., *Contemporary Perspectives on Early Modern Philosophy*. Peterborough, ON: Broadview Press, 2008: 157–71.

Jolley, Nicholas. 'Dull Souls and Beasts: Two anti-Cartesian Polemics in Locke', in J. Hill and P. Glombíček, eds., *Essays on the Concept of Mind in Early-Modern Philosophy*. Newcastle upon Tyne: Cambridge Scholars Publishing, 2010: 94–114.

Kambouchner, Denis. 'Descartes et l'indépendance de l'esprit', *Intellectica*, 57/1 (2012), 55–67.

Kambouchner, Denis. *Descartes n'a pas dit*. Paris: Les Belles Lettres, 2015.

King, Peter, ed. *The Life and Letters of John Locke with Extracts of his Journals and Commoplace Books*. London: Bell and Daldy, 1864.

La Forge, Louis de. *Treatise on the Human Mind*, trans. Desmond M. Clarke. Dordrecht: Kluwer, 1997.

Lähteenmäki, Vili. 'The Sphere of Experience in Locke: The Relation between Reflection, Consciousness and Ideas', *Locke Studies*, 8 (2008), 59–100.

Leibniz, Gottfried Wilhelm. 'Critical Thoughts on the General Part of the Principles of Descartes' (1692) in *Philosophical Papers and Letters*, ed. L. E. Loemker. Dordrecht, Boston, New York: Kluwer Academic Publishers, 1989.

Leibniz, Gottfried Wilhelm. *New Essays on Human Understanding*. Cambridge: Cambridge University Press, 1996.

LoLordo, Antonia. *Locke's Moral Man*. Oxford: Oxford University Press, 2012.

McCann, Edwin. 'Cartesian Selves and Lockean Substances', *The Monist*, 69/3 (1986), 458–82.

Rabb, J. Douglas. *John Locke on Reflection, A Phenomenology Lost*. Washington, DC: Center for Advanced Research in Phenomenology,1985.

Radner, Daisie. 'Thought and Consciousness in Descartes', *Journal of the History of Philosophy*, 36 (1988), 439–52.

Rozemond, Marleen. 'Descartes's Case for Dualism', *Journal of the History of Philosophy*, 33/1 (January 1995), 29–63.

Rozemond, Marleen. *Descartes's Dualism*. Cambridge, Mass.: Harvard University Press, 1998.

Simmons, Alison. 'Cartesian Consciousness Reconsidered', *Philosophers' Imprint*, 12/2 (2012), 1–21.

Simonetta, David. 'L'âme pense-t-elle toujours? Postérité de la théorie de l'*intensio* et *remissio formarum* dans la querelle entre empiristes et cartésiens', *Astérion*, 11 (2013) <http://asterion.revues.org/2434>.

Thiel, Udo. *The Early Modern Subject*. Oxford: Oxford University Press, 2011.

8

Locke and Descartes on Free Will

The Limits of an Antinomy

Denis Kambouchner

There is no doubt that the question of free will affords a prime arena for comparing or contrasting the *Essay Concerning Human Understanding* with the texts of Descartes's 'first philosophy'. The comparison, however, is rendered difficult on two accounts.

First, the problem of freedom cannot be approached in the same way as that of innate ideas or the distinction between the body and the soul (or thinking matter): there is no reason to turn Descartes even implicitly into the official adversary, or one of the official adversaries, targeted by the arguments put forward in the chapter *Of Power* (E II. xxi). These arguments are aimed in a critical manner at a much wider, older and more diversified tradition than the 'new philosophy'; I use the term 'diversified' because it includes both authors who see the will as controlling all the operations of the soul ('*voluntas… actionum nostrarum domina*', as Van Limborch says[1]) and authors who see it as blind and subordinate to understanding. In any event, the structural opposition to Cartesianism that may be found here is far less direct, pressing, and typical than it is on other major themes in the *Essay.*

Moreover, both Locke and Descartes are known to have formulated the problem of free will in a variety of ways.

On this topic, the recent interpretations of Descartes unfailingly take place in the wake of the century-old thesis by Étienne Gilson, *La Liberté chez Descartes et la théologie*, which stressed the discrepancy between the very Thomistic formulations of Meditation 4 and those, seemingly far more Molinist, of the *Principia* of 1644 or, even more so, of the letters to Father Mesland of 1644 and 1645. Gilson attributed this discrepancy not to deeper reflection by Descartes, but rather to pure doctrinal opportunism or a carefully calculated pragmatism, linked primarily to Descartes's desire to win over the Jesuits. In contradiction to Gilson, numerous studies have since attempted to demonstrate the

[1] Van Limborch to Locke, 3/14 October 1701; LC 7, p. 450 (letter 3010).

fundamental unity of the Cartesian position;[2] but it has to be acknowledged that the task is arduous, and confirming this unity means digging deep.

As for Locke, we know how the deterministic wording of the first edition of the *Essay* was toned down or mitigated in the subsequent editions—as early as the second edition, with the new emphasis put on the power to suspend our desires, and once again in the fifth edition, where Locke concedes that there is a case 'wherein a Man is at Liberty in respect of *willing*' (§56). While concern for the reception of the theses contained in the *Essay* was clearly a decisive factor in these modifications, Locke studies have not had their Gilson. There is no reason to suppose any hidden, opportunistic, motive behind these changes, since Locke himself quite openly acknowledges at the end of II. xxi (§72) and in the Epistle to the Reader, that he was somewhat mistaken in what he wrote in the first edition.[3]

Nevertheless, an analogous problem is raised, and this problem—recently tackled with great skill by Richard Glauser[4]—concerns the extent to which Locke's text offers safe ground for the claim that the kind of freedom Locke has come to recognize in man, with respect to his thoughts, his desires, and even his will, is compatible with the assertion of a psychological ('volitional') determinism.

Arguably, the situation is tricky, and before two doctrines, one not clearly addressing the other, can be compared, it may seem that a first requirement would be that each position was stable enough to be distinctly characterized. Personally, however, I am inclined to take the opposite position, namely that:

(1) It is this very evolution in their thinking and formulating of the problem of free will that the two authors have above all in common (if we consider the direction this evolution took).
(2) This evolution is not strictly peculiar to them; a similar evolution can be seen in Malebranche (to all intents and purposes the intermediary here) and to this extent there is, despite the accepted divisions and very real differences in emphasis, a real family resemblance among the three authors.
(3) This family resemblance hangs on their recognition of the irreducibility of the problem, a recognition at one with a language that dismisses overly conventional positions, so that the opposition or antinomy observed prima facie between Locke's and Descartes's doctrines subsequently requires considerable shading, until in the end it is their affinities that stand out.

[2] See esp. Laporte, 'La Liberté selon Descartes'; Gabaude, *Liberté et raison*. vol. 1; Beyssade, *La Philosophie première de Descartes*; Bouchilloux, *La Question de la liberté chez Descartes*; Kambouchner, *Descartes et la philosophie morale*, ch. 1. Lennon, 'Descartes's Supposed Libertarianism', recently questioned the status of the so-called 'letter' to Mesland of 9 February 1645, considering it as a 'memorandum' rather than as a real piece of correspondence. To his eyes, however, this writing is 'perfectly consistent with the compatibilist position taken in the fourth *Meditation*' (p. 229).

[3] See E Epistle to the reader, p. 11, and E II. xxi. 72: 'In the former Edition of this Treatise, I gave an account of my thoughts concerning them, according to the light I then had: And now as a Lover of Truth, and not a Worshipper of my own Doctrines, I own some change of my Opinion, which I think I have discover'd ground for. In what I first writ, I with an unbiassed indifferency followed Truth, whither I thought she led me. But neither being so vain as to fancy Infallibility, nor so disingenuous as to dissemble my mistakes for fear of blemishing my reputation, I have with the same sincere design for truth only, not been asham'd to publish what a severer enquiry has suggested.'

[4] Glauser, 'Thinking and Willing' and Glauser 'Liberté, compatibilisme et agnosticisme chez Locke'.

1

In the profusely worded and complex chapter II. xxi, there are two features that appear to be clearly anti-Cartesian, or at least at odds with Cartesian formulations and the theoretical intentions they express.

(1) To start with, there is the first, most forthright and famous position declared in the *Essay*, virtually highlighted by the subheadings of the chapter: the idea that liberty does not belong to volition (§10), nor to will (§14 and 20). Powers belong only to agents (§16–19) and thus liberty belongs solely to the agent or the man (§21) and, 'in respect of willing', man is not free (§22–4).

This is not the place to examine the detail of Locke's arguments, which are complex both in construction and in scope. Obviously, before the arguments that he uses to deny the freedom of the will, we need to consider the shift entailed in the very definition of liberty given in §8, with respect to its common metaphysical one.

First, this definition exclusively relates liberty (in its true sense) to an action which, according to the examples in §9, is to be considered as mainly (I am not saying exclusively) of an external, transitive, or physical nature (an action which, as Descartes says somewhere, 'passe hors de ce qui agit' ('passes outside what is acting'). Second, and correlatively, this definition makes the true nature of liberty totally subordinate to that of a preference or volition already formed according to a determination of the mind (§8) or of the will.[5] Insofar as evaluating liberty implies referring to a preference or a volition which is already formed, the volition itself cannot be qualified as free: that is, because it is meaningless to assert that our volition is the object of another volition, or even worse, that we *prefer to prefer* one thing to another.

Take the conceptual argument according to which, the will being a power ('*Power* to begin or forbear, continue or end several actions of our minds, and motions of our Bodies', §5) that relates to an agent, and liberty likewise (the power to act or not to act according to one's preference), it is meaningless to attribute liberty to will, because it is meaningless to attribute one power to another (it can only be attributed to an agent). The argument here is clearly tied to this first position, but is not a constituent of it and gives rise to particular difficulties.

Another problem is that of the relationship between the initial double decision, concerning the nature or the true idea of freedom, and the doctrine presented several decades earlier by Hobbes in the course of his controversy with Bramhall.

In a classic article, Vere Chappell pointed out the division between the two definitions.[6] Hobbes speaks of freedom once we can do what we wish, and in this respect, a voluntary action and a free action amount to the same thing. Locke, on the other hand, distinguishes between the free and the voluntary, and maintains within the exercise of freedom the principle of a *potestas ad utrumlibet*, a 'two-way power'.

[5] *Determinatio voluntatis*: Locke to Van Limborch, 21 May 1701; LC 7, p. 328 (letter 2925).

[6] See Chappell, 'Locke on the Freedom of the Will'. On Hobbes, see also Pink, 'Hobbes on Liberty, Action, and Free Will'.

Clearly, Locke always remains careful not to cross the line that separates reasonable psychological determinism from provocative necessitarianism. Nevertheless, he transforms the fundamentals of liberty as much as Hobbes, in his more straightforward manner, had done: he brings liberty down to the empirical and contextual conditions of an action and removes it, at least at first assessment, from the commanding intention itself. Hobbes's definition of the free agent, 'a free agent is he that can do if he will, and forbear if he will',[7] remains valid in a Lockean sense, so to speak, even if the Hobbesian definition of liberty as an 'absence of external impediments' may seem too restrictive and, above all, too negative. I will return to this point.

To the extent that Locke's work is following a certain Hobbesian tradition, it appears clearly anti-Cartesian-inspired. Or rather, Descartes seems to be taking his place in the wide spectrum or the long series of authors, ranging from the Stoics to numerous Moderns via Augustine and Duns Scotus, whose manner of thinking or of speaking Locke dismisses.

While it is true that Descartes attaches at least one meaning of liberty to the external conditions of action, these external conditions define but a secondary dimension of liberty, the primary dimension being related to the determination of the will itself. This freedom of the will is asserted as a fundamental principle in several texts. At this junction, I will cite from among the Latin texts only the *Principia* I, art. 39:

> *Quod autem sit in nostra voluntate libertas, et multis ad arbitrium vel assentiri vel non assentiri possimus, adeo manifestum est, ut inter primas et maxime communes notiones, quae nobis sunt innatae, sit recensendum.*
>
> That there is freedom in our will, and that we have power in many cases to give or withhold our assent at will, is so evident that it must be counted among the first and most common notions that are innate in us.[8]

Several statements derive from this axiom in *The Passions of the Soul*, one on the nature of the will ('The will is by its nature free in such a way that it can never be constrained'[9]) and the others on the freedom of the will as an object of the 'generous' thinking:

> I observe but a single thing in us which could give us just cause to esteem ourselves, namely, the use of our free will and the dominion we have over our volitions.[10]
>
> True generosity, which makes a man esteem himself as highly as he can legitimately esteem himself, consists only in this: partly in his understanding that there is nothing which truly belongs to him but this free control of his volitions...[11]

[7] *Of Liberty and Necessity*, in *The English Works of Thomas Hobbes*, vol. 4, 275.

[8] AT 8A, p. 19: CSM 1, pp. 205–6.

[9] 'La volonté est tellement libre de sa nature, qu'elle ne peut jamais être contrainte': art. 41; translation by Stephen Voss, Descartes, *The Passions of the Soul* (Indianapolis/Cambridge: Hackett, 1989) (hereafter Voss), p. 41.

[10] 'Je ne remarque en nous qu'une seule chose qui nous puisse donner juste raison de nous estimer, à savoir l'usage de notre libre arbitre, et l'empire que nous avons sur nos volontés': art. 152, Voss p. 103.

[11] 'La vraie générosité, qui fait qu'un homme s'estime au plus haut point qu'il se peut légitimement estimer, consiste seulement, partie en ce qu'il connaît qu'il n'y a rien qui véritablement lui appartienne, que cette libre disposition de ses volontés...': art. 153; Voss p. 104.

and elsewhere:

Generosity... makes us greatly esteem liberty and absolute dominion over ourselves...[12]

For the moment, I will leave aside the famous letter to Father Mesland of 9 February 1645, which takes free will or free choice to its ultimate limits by talking of a choice made in contradiction with a clear knowledge of truth or good.

(2) This brings us to the second aspect. Not only does Locke bring liberty down from the position of determining action to that of action itself viewed as it is triggered or begins, but he also accepts that the preference or volition the action springs from (insofar as it is within our power) is determined by something (by an internal state of mind). Coste's French translation could give the opposite impression:

chaque Homme est... dans la nécessité, en vertu de sa propre constitution... de *se déterminer* à vouloir ce que ses propres pensées & son jugement lui représentent pour lors comme la meilleure chose qu'il puisse faire. (EEH II. xxi. 48, p. 205) (my italics)

But this is because a secret Cartesian sympathy has misleadingly transformed a passive structure in Locke's text:

every Man is put under a necessity by his constitution... to *be determined* in willing by his own Thought and Judgment, what is best for him to do (my italics)

into a French reflexive structure, although it is clearly the passive that is intended here, as well as in the other Lockean statements.

On this point, whatever the complications introduced by the passages added to the later editions, the deterministic wording of the first edition (highlighted, here too, by the subheadings) was never removed.[13] Although, in his second edition Epistle to the Reader, Locke speaks of the reasons that have led him to revise 'the thoughts I formerly had concerning that, which gives the last determination to the Will in all voluntary actions' (p. 11), he sticks firmly to the principle that the will is determined *by something else*. And while this very text evokes a determination operated *by the will*, it is not the will itself but our power to act that is being determined. In this connection, I will later cite a text where Locke literally accepts a degree of self-determination, although this is far removed from the status given it by someone like Cudworth, for example.

Now, this way of speaking of the will or of volition as being determined by something else is in total opposition with the Cartesian account. The Cartesian assertion of the freedom of the will, which assumes axiomatic proportions, basically implies that the will cannot be determined by anything external to itself. An expression such as 'the will is determined' is nowhere to be found in Descartes—neither in Latin, nor in

[12] 'La vraie générosité, faisant... qu'on estime beaucoup la liberté et l'empire absolu sur soi-même...': art. 203; Voss p. 129.

[13] It is even reinforced by the 1694 corrections: see Pécharman, '*Il faut parier*: Locke ou Pascal?', pp. 490–1.

French. It would mean not only that the soul is bound to want what it wants, but that even its will is subject to a kind of constraint. On the contrary, Descartes always maintained the opposing Stoic-Augustinian position, never challenged since it stands as a fundamental principle, which is that, in truth, our will always belongs to itself. Thus, when great pressure is brought to bear on it—either through blinding evidence, or on a very different level, through a very violent passion—it is always the will itself that *surrenders*, either accepting the evidence or giving in to the pleas or 'efforts' of the passion. Evidence of this position can easily be found in a whole set of statements, in the *Meditations* and the *Replies to Objections*, as well as in *The Passions of the Soul*. The only way for this self-determination to be eliminated is for the will as such to be eliminated or to disappear, which is what happens in certain illnesses 'which take away the power of reasoning',[14] in somnambulism or similar cases.

Should we also consider as anti-Cartesian the view expressed in the *Essay* that it is not what is perceived as the greatest good at the time which ordinarily determines the will, but rather 'the most important and urgent uneasiness, we at that time feel' (II. xxi. 40)? I do not believe so, but I will discuss this point in Section 3.

2

The opposition or antinomy observed here between the two authors thus appears to be very pronounced. However, closer examination is required, on each of the two different levels I have identified.

(1) First let us consider, starting with Locke, the definition of freedom. While Locke initially rejected any mention of freedom for willing itself, the fact is that successive editions of the treatise attenuated this rejection. This is, and would remain, understandable even were Locke's position later in the *Essay* (and in the subsequent editions) not subject to the overriding influence of moral considerations.

That the true notion of liberty, insofar as it involves a current preference or volition, makes the very idea of freedom for volition meaningless—this is what Locke starts off by saying and repeating. Despite the fact that the question of whether man's will is free or not has been debated for a very long time, it is, as he famously points out in §14, 'unreasonable, because unintelligible', or in other words 'altogether improper' (see also §21). In reality, says §25, 'Liberty concerns not the Will'.

According to this doctrinal line, we can legitimately *neither* affirm that our volitions are free *nor* deny that they are free. However, this is a difficult position to hold: difficult for the generality of men, always seeking a maximum of freedom (§22: 'The inquisitive Mind of Man…is not content with this'); but difficult too, it would seem, for Locke himself, who appears immediately to transgress his own rule by undertaking to show that, 'in respect of willing, a Man is not free' (§§22–4). It is all as if Locke's initial

[14] Descartes to Elisabeth, 1 September 1645, AT 4, p. 282; CSMK, p. 262.

definition were, as Kant was to say about the theses of the antinomies of pure reason, 'too small for our understanding'; and this is in some ways what Van Limborch later reproached him with: 'your definition seems to me to be too narrow (*angusta nimis*)'.[15]

The strange thing is that transgression of the rule: 'Liberty concerns not the Will' operates in two different directions. On the one hand, we move from indeterminability towards negation, in other words, from *we cannot say that the will is free,* or *we cannot tell whether the will is free,* to *the will is not free*; on the other hand, we are back with the question of some indirect power that we have over our volitions.

The latter transgression was predictable, since Locke began by holding that our power extends not only to our external or physical actions, but also to some extent to our thoughts; see §8:

> All the Actions, that we have any *Idea* of, reducing themselves, as has been said, to these two, *viz.* Thinking and Motion, so far as a Man has a power to think, or not to think; to move, or not to move, according to the preference or direction of his own mind, so far is a Man *Free*.

and §12:

> As it is in the motions of the Body, so it is in the Thoughts of our Minds; where any one is such, that we have power to take it up, or lay it by, according to the preference of the Mind, there we are *at liberty*.

Saying *thoughts* is not the same thing as saying *volitions*; yet the possibility remains that thoughts which are not volitions may determine a volition; in other words that the mind, to which the freedom to desire cannot be directly attributed, may come by it indirectly, through the sort of power that he exercises over his other thoughts. The extent to which he exercises this power and examines things in the proper manner will determine the extent to which we can say that his will is not determined by anything other than 'his own *desire* guided by his own *Judgment*' (§71).

Nowhere, it is true, is Locke seen in this discussion to return to the concept he is trying to detach himself from, that of a supreme power governing the use of the other faculties and even its own use. There is no double-talk from him in this respect. The expression 'free will' itself appears nowhere without express reservations (see §47 giving an object to 'that which is (as I think improperly) call'd *Free will*'). As for self-determination, it is relegated to the fringes, as shown by a fragment from 1694, which was included in a letter to Jean Le Clerc on 9 October 1694, although it was not added to the *Essay*'s third edition in 1695:

> A great number of litle and very indifferent actions which mix themselves with those of greater moment and fill up as it were the little empty spaces of our time. In these *the will may be said to determin itself* without the preponderancy of good or evil or the motive of uneasinesse on either side.[16]

[15] Van Limborch to Locke, Letter of 3 October 1701, LC 7, p. 458 (letter 3010).

[16] MS Locke c. 28, fo. 115^{v} (my emphasis). See LC 5, p. 159. Locke continues: '...as whether a man should put on his right or left shoe first. Whether he should fold a margent or noe in the paper wherein he

The fact remains that under certain conditions, admittedly not the most frequent, the power to examine which reappears in §46 and following articles amounts to restoring a certain freedom of choice. This is perhaps conceded a lot earlier in the chapter than would appear (§23), but I will return to this point.

(2) On Descartes's side, notwithstanding the blatantly clear formulations quoted earlier, it needs to be recognized that, on the question of the freedom of the will, there is a much greater reserve or prudence than is generally believed. First of all, Descartes does not have much to say about liberty. In his metaphysical and his moral texts alike, the word occurs amazingly rarely as a real theme. The common rhetoric on the free will is, moreover, used very sparingly in his texts. And a cautious look at the texts will show that the famous theme of the 'infinity of the human will' is a quasi-artefact, no doubt with a complex historiography, but constructed on the thinnest of textual bases.[17]

Furthermore, in certain formulations of a moral nature, 'the free control of our volitions', in other words, 'the absolute dominion over ourselves', is obviously not considered as a pure ideal: everyone must be able to experience this control and to some extent achieve it within himself. Conversely, however, it can never be a general disposition capable of being fully testified by a certain form of experience; rather, it is always the object of an exercise.

As for the freedom of the will in general, Descartes treats it in no way as the power of another power, product of a purely speculative psychology, but as the common denomination of a basically irrefutable experience. This is what is behind the replies to Hobbes (and also to Gassendi): *Nihil autem de libertate hic assumpsi, nisi quod omnes experimur in nobis*, 'Regarding freedom I made no assumptions here, except what we all experience within ourselves.'[18] And as Descartes stresses both to Hobbes and in Meditation 4 using the equivalence: *facultas eligendi, sive voluntas, sive libertas arbitrii*, the experience of freedom and that of willing are absolutely one and the same thing ('voluntariness and freedom are one and the same thing'[19]). It follows that this experience is also indivisible, and to anyone who denies the freedom of the will (Gassendi), the reply is: *Ne sis igitur libera [ô caro], si non lubet; ego certe mea libertate gaudebo, cum et illam apud me experiar, et a te nulla ratione, sed nudis tantum negationibus impugnetur*,

is goeing to write a letter to his friend, whether he should sit still or walke or scratch his head whilst he is in a deep meditation there are a thousand such actions as these which we doe every day which are certainly voluntary and may be ascribed to the will determining it self, but there is soe litle thought preceeds them because of the litle consequence that attends them that they are but as it were appendixes to the more weighty and more voluntary actions to which the minde is determind by some sensible uneasinesse and therefor in these the minde is determind to one or tother side not by the preferable or greater good it sees in either, but by the desire and necessity of dispatch that it may not be hinderd in the pursuit of what is judgd of more moment by a lingring suspense between equall and indifferent things and by a deliberation about trifles in these the uneasinesse of delay is sufficient to determin and give the preference to one it matters not which side.'

[17] I developed this point in *Descartes n'a pas dit*, ch. 10, pp. 101–9.

[18] Third Set of Replies, 12, AT 7, p. 191; CSM 2, p. 134 (modified).

[19] 'Le libre et le volontaire n'étant en effet qu'une même chose...': *Third Set of Objections with Replies*, 12th Objection, AT 7, p. 191; CSM 2, p. 134.

'You may be unfree, if you wish; but I am certainly very pleased with my freedom since I experience it within myself. What is more, you have produced no arguments to attack it but merely bald denials.'[20]

Only two points need, I believe, to be underlined here. First, as Meditation 4 makes quite clear, the defining experience of the free will can be found in the search for truth, when things that appeared absolutely manifest are placed in doubt. This is what the *Synopsis of the Meditations* says: *mens quæ, propria libertate utens, supponit ea omnia non existere de quorum existentia vel minimum potest dubitare, animadvertit*.... 'the mind, using its own freedom and supposing the non-existence of all the things about whose existence it can have even the slightest doubt, notices, etc.'[21]

Thus, what is first experienced is the freedom that we have to suspend our judgement, or assent—freedom to doubt, that makes so valuable the later acceptance of certain facts or propositions recognized as self-evident. This freedom, as presented by Descartes, obviously has no Lockean equivalent. But Locke, too, is eventually forced to recognize a certain freedom to suspend, first as a 'power (of the mind) to *suspend* the prosecution of this or that desire' (§47), but later as a power over the desire itself, a power which indeed constitutes 'that which is (as I think improperly) call'd *Free will*'. What is important is that Locke accepts that there is actual experience of this power. And if this experience only takes place against the background of competing anxieties that the mind cannot do more than record, there are Cartesian precedents for rooting the power to suspend in a state of mind composed initially in a non-intentional manner. In the first of his two letters on liberty to Mesland (2 May 1644), a letter less famous than the second (9 February 1645), Descartes writes:

> I agree with you when you say that we can suspend our judgment; but I tried to explain in what manner this can be done. For it seems to me certain that, *ex magna luce in intellectu sequitur magna propensio in voluntate*;[22] so that if we see very clearly that a thing is good for us, it is very difficult—and, on my view, impossible, as long as one continues in the same thought—to stop the course of our desire. But the nature of the soul is such that it hardly attends for more than a moment to a single thing; hence, as soon as our attention turns from the reasons which show us that the thing is good for us, and we merely keep in our memory the thought that it appeared desirable to us, we can call up before our mind some other reason to make us doubt it, and so suspend our judgment, and perhaps even form a contrary judgment.[23]

The paradox is intentional: this distracting of attention as the only means of suspending judgement with respect to a very clear representation of a good is not presented as a purposeful act of will, but rather as something unavoidable because 'the nature of the soul is such that it hardly attends for more than a moment to a single thing'.

[20] Fifth Set of Replies, IV, 3, AT 7, p. 377; CSM 2, p. 259.

[21] AT 7, p. 12; CSM 2, p. 9 (translation modified).

[22] 'A great light in the intellect is followed by a great inclination in the will.' From Meditation 4, AT 7, p. 59; CSM 2, p. 51 (translation modified).

[23] Descartes to Mesland, 2 May 1644, AT 4, pp. 115–16; CSMK, pp. 233–4.

3

The crux of the opposition between Locke and Descartes can perhaps be found in volitional or psychological determinism. Here again, however, certain complexities should be noted.

(1) On Descartes's side first: it is obvious that he is taking numerous grammatical precautions with his wording; these precautions are integral parts of his highly disciplined way of thinking and writing, and are quite striking, for example, when it comes to the infinite power of God (since 'we should [n]ever say of anything that it cannot be brought about by God',[24] it is impossible to say that 'God cannot do this', it can, however, be said that 'God cannot be...' or 'it cannot be that God will do this').[25] As for the will, when we say that it is determined by something, we are making a statement contrary to its very nature; which is why Descartes urges Gassendi to consider 'the freedom (that the will has) to direct itself, without the determination of the intellect (*sine determinatione intellectus)*, towards one side or the other'.[26] In practice, nonetheless, we have to recognize that the will is driven in a variety of ways, that it never determines itself without motive and that, 'morally speaking', *moraliter loquendo* (as the second letter to Mesland states it), it never fails to surrender to certain types of perception. This was stipulated, for example, in the *Exposé géométrique* following the *Second Set of Replies,* in the 7th axiom or common notion:

Rei cogitantis voluntas fertur, voluntarie quidem et libere (hoc enim est de essentia voluntatis), sed nihilominus infallibiliter, in bonum sibi clare cognitum.

The will of a thinking thing draws itself voluntarily and freely (for this is the essence of will), but nevertheless inevitably, towards a clearly known good.[27]

'The will.,.. draws itself voluntarily' means that the will itself agrees to draw towards something; but here, of course, as elsewhere, this will is not merely a faculty: it is simply the soul itself currently acting on a certain level.

But to return to the main point: it is one thing to say that the will cannot be constrained, and another thing to say that it can always manage to evade the attraction of certain perceptions, and yet another thing to say that its determination knows absolutely no rules. And if Locke protests (§50) against the idea of a liberty that consists of shrugging off the yoke of reason and not bothering to examine things, Descartes's condemnation of this notion of liberty is no less firm. The letter to Mesland of 9 February 1645 even goes so far as to speak of 'the positive power which we have of following the

[24] 'On ne doit jamais dire d'aucune chose qu'elle est impossible à Dieu': Descartes to Arnauld, 29 July 1648, AT 7, p. 224; CSMK, p. 358.

[25] These distinctions were clearly made by Jean-Marie Beyssade; see esp. *La Philosophie première de Descartes*, pp. 115–21.

[26] AT 7, p. 378; CSM 2, p. 260.

[27] Second Set of Replies, AT 7, p. 166; CSM 2, p. 117. We have corrected the English translation, which unduly uses the passive ('is drawn voluntarily...').

worse although we see the better'[28] and the fact that 'it is always open to us to hold back from pursuing a clearly known good, or from admitting a clearly perceived truth, provided we consider it a good thing to demonstrate the freedom of our will by so doing.'[29] First, however, it is easy to consider, like Leibniz, that given such a 'whim,'[30] attesting to the free will seems a greater good than the good we are turning away from (or than the truth we are rejecting): thus, we might well agree with Locke that the most pressing uneasiness is that of the attendant desire. And secondly, while in 'absolute terms' the exercise of this freedom can be imagined, in 'moral terms' it can only be exceptional and at the agent's own risk.

Furthermore, when Descartes discusses how the will interacts with present passions, which tend to take hold of the imagination and 'always make both the goods and the evils they represent appear much greater and more important than they are,'[31] it turns out that the will is not able to resist these passions with ease, except if it is firmly attached to a form of good that it considers to be even greater than any of those which these passions represent. I am thinking of articles 48 and 49 of the *Passions de l'âme*:

> Those in whom the will can naturally conquer the passions most easily and stop the accompanying movements of the body have the strongest souls. However, there are some who cannot test their strength, because they never make their will do battle with its proper weapons... What I call its proper weapons are firm and decisive judgments concerning the knowledge of good and evil, which it has resolved to follow in conducting the actions of its life.... The greater part [of men] have decisive judgments which they follow in regulating a part of their actions. And though these judgments are often false, and even founded on passions by which the will has previously allowed itself to be conquered or seduced, because it continues to follow them when the passion that caused them is absent, they can be regarded as its proper weapons, and souls can be thought to be stronger or weaker to the extent that they are more or less able to follow these judgments and resist the present passions opposed to them.[32]

(2) On Locke's side, the question of the determination of the will is far from being as simple as declared. If it were, in fact, each volition could or should be called necessary, and Van Limborch would not be quite wrong in suggesting[33] that this would put an end to freedom. Even before the second edition, the difficulty is clear in §23, expressed in what strikes us as extraordinary terms, considering its heading (*In respect of willing, a man is not free*). What Locke actually says is that once an action is suggested to a man's mind as needing to be performed immediately, that man is no longer free to desire or not to desire; that it is no longer in his power to refrain from desiring or from not desiring (this point is recalled in §56). But this apparently means no more than that everyone must, under these conditions, choose OR not choose that which is presented to him, *prefer the doing or forbearance (of this action which is in his power).*

[28] Descartes to Mesland, 9 February 1645, AT 4, p. 174; CSMK, p. 245.
[29] Descartes to Mesland, 9 February 1645, AT 4, p. 173; CSMK, p. 245.
[30] Leibniz, *Essais de Théodicée*, §45.
[31] *The Passions of the Soul*, art. 138, AT 11, p. 431: Voss p. 93.
[32] AT 11, pp. 366–8, Voss pp. 46–7.
[33] Van Limborch to Locke, 3/14 October 1701; LC 7, p. 463 (letter 3010).

As Coste translates it in §24, '*[cet homme] n'est plus en liberté de vouloir vouloir*', that is, the man is obliged to desire something: the very suggestion of this action to his thoughts means that an effect must necessarily follow on the part of his will, and that it is metaphysically impossible for him not to determine himself in one direction or the other.

But the objection can be raised that, if we accept that we have to determine ourselves, freedom actually consists for the main part in how we determine ourselves; and Locke here seems to be illustrating the opposite of what he intends, for there is no question here of prescribing the manner in which the mind takes decisions on what is suggested to it, and we are not talking about the case, which he recognizes as frequent, where the uneasiness of a certain desire directly conditions volition, or 'immediately determines' it (§33).

Locke's feeling may be that, under these conditions, it is morally impossible for the will not to focus on the suggested action: this would mean that our mind stands in relation to this suggestion like the man locked in a room with a person that 'he longs to see and speak with' (§10), and who could not possibly fail to speak with this person even if not physically constrained to do so.[34] No doubt the fact is that we cannot set ourselves to willing, and that we are always incited to it: we may will one thing or another, but we are bound to will something. And yet we get the distinct impression here that to be determined to such volition is not the same as to be determined to a choice in general, and that the former determination possesses, so to speak, its own economy. Indeed, if the rest of the chapter is concerned with introducing, through the theory of uneasiness, a sort of law governing the determination of our volitions, Locke scarcely sheds any light on the precise mode of this determination. We are, says Locke, determined *by our last judgement;* see §71: 'The result of our judgment upon that Examination is what ultimately determines the man.' And §48: 'Nay were we determined by any thing but the last result of our own Minds, judging of the good or evil of any action, we were not free.' Yet how we arrive at this last judgement seems difficult to explain clearly, and while Locke speaks of a series of consequences, it occurs after, rather than before, this judgement:

> What follows after that [examination of the good and evil of what we desire], follows in a chain of Consequences linked one to another, all depending on the last determination of the Judgment. (§52)

But this, adds Locke, 'whether it shall be upon an hasty and precipitate view, or upon a due and mature *Examination*, is in our power.'

4

In short, what unites Descartes and Locke can be said here to be an extremely acute awareness, no doubt heightened by the objections, of the extreme complexity of the

[34] This can be compared with the example of the duels in the letter to Elisabeth of January 1646, AT 4, pp. 353–4; CSMK 3, p. 282.

question (of the conditions under which our volitions are determined). With Descartes, this complexity is clearly shown in his letter to Mesland of 9 February 1645; and with Locke in the sections 47 and following.

In a recent study,[35] Richard Glauser examined the extent of Locke's agnosticism, which was very marked on several points (the action of the mind on the body, the compatibility of divine omniscient power with human liberty): does this agnosticism extend to psychological determinism, or indeed can Locke be said to be taking a firmly compatibilist position on the relationship between human liberty and psychological determinism? I do not intend to enter into the finer points of this debate. What I feel is key here is that while, on the one hand, Locke considers the unavoidable determination of volitions within a certain economy to be an absolutely valid philosophical principle, on the other hand, as a reader of Malebranche, he cannot fail to be aware that it is impossible to determine exactly or entirely *what happens within us* when we think or when we desire. With Locke, recognizing a form of free will—specifically linked to cases where the mind veers back and forth among several different sides or reasons, perhaps among several types of good, and is led to reflect on its own determination—coincides with recognizing that it is impossible to unravel the whole process of deliberation, to obtain a clear view of how our thoughts determine one another and the time-frame within which they do so. This amounts to saying that a necessitarian view of human action is basically impossible to *achieve*.

In this respect, give or take a few sign changes, Locke and Descartes are on the same moderate side, one as far removed from the exaltation of human freedom as it is from the provocations of naturalism. Leaving aside certain epistemological features—like calling on inner experience or reflection in order to grasp the nature of freedom—or certain metaphysical features—like abandoning the attempt to unravel the relationship between man's freedom and divine providence—there are numerous aspects of their moral theory which could demonstrate their affinity: for instance, the distinction they make between cases of greater or lesser freedom; the distinction they make between the situations which allow an examination to take place and those which do not; how they prescribe the closest examination of the goods sought; the idea that there is a taste of the mind which can be cultivated, and the idea of how it should be cultivated.

A crucial question now appears to be that of the role Malebranche played in the construction of this affinity.

(Translation: Marjorie Sweetko)

References

MS Locke c. 28 Oxford, Bodleian Library. The Locke Digital Project <http://www.digitallockeproject.nl>.

Beyssade, Jean-Marie. *La Philosophie première de Descartes*. Paris: Flammarion, 1979.

[35] 'Liberté, compatibilisme et agnosticisme chez Locke', pp. 682–95.

Bouchilloux, Hélène. *La Question de la liberté chez Descartes*. Paris: Champion, 2003.

Chappell, Vere. 'Locke on the Freedom of the Will', in G. A. J. Rogers, ed., *Locke's Philosophy: Content and Context*. Oxford: Oxford University Press, 1995.

Descartes, René. *The Passions of the Soul*, trans. Stephen Voss. Indianapolis/Cambridge: Hackett, 1989.

Gabaude, Jean-Marc. *Liberté et raison: la liberté cartésienne et sa réfraction chez Spinoza et chez Leibniz*. Vol. 1. Toulouse: Presses de la Faculté des Lettres de Toulouse, 1970.

Gilson, Étienne. *La Liberté chez Descartes et la théologie*. Paris: Alcan, 1913.

Glauser, Richard. 'Thinking and Willing in Locke's Theory of Human Freedom', *Dialogue*, 42/4 (2003), 695–724.

Glauser, Richard. 'Liberté, compatibilisme et agnosticisme chez Locke', *Revue Philosophique de Louvain*, 4 (2009), 675–97.

Hobbes, Thomas. *The English Works of Thomas Hobbes*, ed. W. Molesworth, London, 1840.

Kambouchner, Denis. *Descartes et la philosophie morale*. Paris: Hermann, 2008.

Kambouchner, Denis. *Descartes n'a pas dit*. Paris: Les Belles-Lettres, 2015.

Laporte, Jean. 'La Liberté selon Descartes', *Revue de métaphysique et de morale*, 44 (1937), 101–64, included in *Études d'histoire de la philosophie française au XVIIe siècle*. Paris: Vrin, 1951: 37–87.

Leibniz, Gottfried Wilhelm. *Essais de Théodicée*, 1710.

Lennon, Thomas M. 'Descartes's Supposed Libertarianism: *Letter to Mesland* or *Memorandum concerning Petau?*', *Journal of the History of Philosophy*, 51/2 (2013), 223–48.

Pécharman, Martine, '*Il faut parier* : Locke ou Pascal?', *Les Études philosophiques*, 95/4 (2010), 479–516.

Pink, Thomas, 'Hobbes on Liberty, Action, and Free Will', in A. P. Martinich and K. Hoekstra, eds., *The Oxford Handbook of Hobbes*. Oxford: Oxford University Press, 2016: 171–94.

9

Essential Religiosity in Descartes and Locke

Catherine Wilson

Both Descartes and Locke were close to the evolving experimental and observational sciences of the seventeenth century, especially as these concerned the nature of human beings. Both were corpuscularian philosophers. Neither one was popular with mainstream Catholic and Protestant theologians. Although both invoked God at critical points in their philosophies, they were striving for a minimal, rational and reasonable, theology. The model of scientific practice and achievement presented to both an alternative to the rude and rough battles of the theologians, and the disunity of the churches enabled them to reject any handmaidenly role for philosophy vis-à-vis theology.[1] Locke declared himself rather an underlabourer to physicists, chemists, and physicians. Descartes thought of his accessory role in grander terms: as supplying conceptual foundations.

There, however, the similarities seem to end. Certain of the central doctrines of Locke's philosophy seem to have been formulated in deliberate opposition to Cartesianism. For Descartes, the laws of nature are simple and few, and at least the 'possible nature' of all phenomena can be and ought to be rationally accounted for by reference to them. Even without knowing exactly how things work, we have considerable power to improve medicine and mechanics with the help of physics (AT 9b, p. 327; CSM 1, p. 289). For Locke, the laws of nature are mostly unknown; we can understand very little of the material world and our ability to make improvements is accordingly limited. Our scientific ignorance is eloquently described in *Essay* IV. iii. 22–30. 'The things that, as far as our Observation reaches, we constantly find to proceed regularly, we may conclude, do act by a Law set them; but yet by a Law, that we know not: whereby, though Causes Work steadily, and Effects constantly flow from them, yet their *Connexions* and *Dependencies* being not discoverable in our Ideas, we can have but an experimental Knowledge of them' (E IV. iii. 29). For Descartes, morality is a matter of custom, convention, and self-interest; for Locke, morality, unlike natural philosophy, is universal,

[1] On science and civility, see Shapin, *Social History*, esp. pp. 114–16.

demonstrable, and divine. Descartes thought he had proved the existence of an incorporeal, separable, and likely immortal soul (AT 7, p. 78; CSM 2, p. 54). Locke appeared to reject, or at least to have serious doubts, about the existence of such an object (E IV. iii. 6). For Descartes, immortality implied the persistence of non-imagistic 'intellectual memory' alone (AT 3, p. 599; CSMK, p. 216); for Locke, immortality implied the restoration of all of one's memories for events in life. Descartes devolved several interesting proofs of the existence of God from the innate idea of God (AT 7, pp. 40–51); Locke denied the existence of all innate ideas, including the idea of God (I. iv. 8). Descartes thought that knowledge worth seeking was not acquired via sense experience; Locke found sense experience and reflection upon it to be the basis of all our useful knowledge. Their views on the will, powers, and causality differed significantly. Leibniz surely had it right when, in the *New Essays*, he typed Locke as 'pretty much in agreement with M. Gassendi' (*New Essays*, VI. vi. 70), Descartes's fiercest and most capable philosophical antagonist.

In a celebrated essay, Richard Ashcraft argued that Locke's religious commitments both 'underlie and vitiate', as Michael Ayers summarized the charge, his general epistemology. Ayers sought to relieve Locke of this charge, and one of my aims is to revisit this controversy. I agree with Ashcraft that Locke needed the Christian religion to be true while at the same time his epistemology pulled him towards religious scepticism and that this set up a profound conflict especially noticeable in the *Essay*. But I agree with Ayers that there is a certain consistency in Locke's epistemology, and I try to bring this out by pointing to the theme of nescience in moral matters that accompanies Locke's minimalistic acceptance of Christianity. The corresponding tension in Descartes is between his constant invocation of God as a transcendental entity juxtaposed against the evident replaceability of the term 'God' with the term 'Nature'. The immutability, inscrutability, and implacability of God in his system implies a certain helplessness in human beings, so that there is a form of 'Cartesian humility', different from 'Lockean humility', but equally interesting for the commentator.

To develop these points, I'll begin with a brief review of the basic theological posits of both philosophers.

1

The Cartesian God as a minimalist object playing a supporting role in the articulation of a scientific worldview has been discussed so extensively that I do not need to go into it very far. Suffice it to say that Cartesian theology embraces four principal points, two concerned with creative power, two concerned with benevolence:

God is the author of the eternal truths (AT 1, p. 145; CSMK, p. 23).

God is the author of nature, i.e. creator of matter and the efficacious legislator of the laws of nature and the ground of causality (AT 11, pp. 32–5; CSM 1, pp. 90–1).

God is the guarantor of human bio-medical capability, though not infallibility (AT 7, pp. 82–3; CSM 2, pp. 56–8).

God is the guarantor of human epistemic competence, though not infallibility (AT 7, p. 54; CSM 2, pp. 37–8).

The power and intelligence of God are displayed in his (one is tempted to say here 'its') originating powers and in the laws of nature, but God's goodness can be appreciated only in the adaptation of the animal body and its sensory systems to life and in the accessibility on the part of the human mind to world-transforming, as opposed to survival-related, knowledge, through controlled rational conjecture, observation, and modelling. Positive or revealed theology does not readily attach to this minimal version, which excludes miracles, can make no real sense of transubstantiation or the Trinity, and whose account of generation leaves no room for Original Sin and its transmission.[2]

Descartes's theology is relatively straightforward and coherent. There is room for philosophical debate about whether the three arguments for the existence of God in Meditation 3, not to mention the recasting of the ontological argument in 5, are any good, or any good in the context in which they appear. There is also room for debate about whether Descartes is an occasionalist or continuous creationist, insofar as evidence seems to exist on both sides. There is further room for debate over whether God's voluntaristic creation of the eternal truths is compatible with the epistemic realism Descartes seems to espouse and with the dependability of human logic and inference. Could an evil demon controlling a purely material universe induce in Descartes, if not the mistaken intuition of his existence, at any rate all the later reasonings of the *Meditations*, thereby deceiving him profoundly? But—fine points aside—the picture hangs together very well.

Descartes's separation of theology from the Christian revelation and Christian sensibility as well as his 'Mechanick Theist' cosmology and theory of nature was responsible for the repudiation of Cartesianism by the English philosophers Henry More and Ralph Cudworth.[3] Newton's enthusiastic defence of God as Creator and Designer, as the foundation of universal attraction, and as a person in the General Scholium of his *Principia* was as significant an instance of English anti-Cartesianism as his repudiation of vortices and hypotheses.[4] For Descartes, the supposition that the world had fallen into its present state through the operations of time and chance (AT 8A, pp. 102–3; CSM 1, pp. 257–8) was entirely coherent and informative, though contrary to Scripture; he exhibited no serious interest in the afterlife, the big promise of Christian revelation. His proof of the distinctness of mind and body was, he insisted, the answer to the challenge presented by the Fifth Lateran Council to philosophers to refute the claim of the soul's mortality (AT 7, p. 3; CSM 2, p. 4). But disembodied minds cannot, according

[2] Nadler, 'Arnauld, Descartes, and Transubstantiation'.
[3] Sailor, 'Cudworth and Descartes'.
[4] On the metaphysical physics of the two, see Janiak, 'Metaphysics and Natural Philosophy'.

to the well-articulated details of his robustly causal theory of perception, have experiences, unless God is a deceiver. Further, insofar as experiences are for preserving biological life, they would have no purpose in heaven.

2

Locke's broad conception seems in turn to acknowledge four roles for God:

God is the all powerful creator of the universe (E IV. x. 5).
God is responsible for the putative superaddition of thought to matter (E IV. x. 10).
Although God is not involved in any acts of perception, God is the author of human beings considered as understanding, rational, dutiful (E IV. iii. 18), and perceptual beings.
God sees into the heart and delivers ultimate justice (E II. xxvii. 22).

In the *Essay*, Locke divided his discussion of God into a section on our idea of God and a section on the existence of God. But rather than invoking causal processes originating with an existing God, as Descartes did in Meditation 3 to explain the occurrence of an idea of God, Locke explained the idea of God as a mental construction. In Draft A of the *Essay*, he seemed to echo Gassendi in the Fifth set of Objections, whose criticism of Descartes's theology ran as follows:

First, Gassendi asked whether Descartes's ascription of the properties of being 'supreme, eternal, infinite, omnipotent and the creator of all things' to his idea of (a so far unproved) God was not simply the product of the tradition. 'Do you not take this from your previously conceived knowledge of God, that is from having heard these attributes ascribed to him? If you had not previously heard anything of this sort would you still describe God in this way?' (AT 7, p. 286; CSM 2, p. 199).

Next, Gassendi questioned whether Descartes could really conceive of an infinite thing and he proposed that the idea of God was constructed by the 'amplification' of 'things we commonly admire in ourselves', such as longevity, power, knowledge, goodness, and blessedness (AT 7, p. 287; CSM 2, p. 200). '[T]ell me in good faith,' he continued a few pages later, 'whether you do not in fact derive all the language which you use of God from the human society in which you live... [A]lthough these ideas do not come from you alone, it seems they do not therefore come from God, but that they come from another source' (AT 7, p. 295; CSM 2, p. 205).

In Draft A of the *Essay*, Locke followed Gassendi in supposing,

> [E]ven the best notions or Idea we can have of god is but attributeing the same simple Ideas of thinkeing, knowing willing existence without beginning & all those powers and operations we finde in ourselves... & conceive to have more perfection in them then [*sic*] would be in their absence, to him in an higher & unlimited degree. (Draft A, #2, 10)

But Gassendi's stunning proposal to Descartes—that the idea of God may be a social construction, even a myth or folktale—was conspicuously not advanced by Locke.

In Draft A, by contrast, Locke went on to say, rather startlingly, that 'the being of a God may be made out more clearly & with greater assureance then of anything even immediately observeable by our senses'. The topic, he promised, would be discussed later 'when I come to treat of things themselves & not of the ways & limits of our understanding' (Draft A, #38, 66).

The admission that we have no positive idea of infinity—that it is simply extension or number considered without limits—will not at all, Locke continued, 'destroy that way by which I suppose we can attein all the knowledg our narrow understandings can arrive to'.

> Nor have we any reason from the narrownesse of our own thoughts to doubt of the Existence of the great god or first aeternall cause because our dim understandings cannot discover his incomprehensible way of being. This were to deny the existence of anything that were too big for our pygmy graspe & boldly to expect althings should be made proportionably to our span. (Draft A, #45, 82)

This is quite a statement, insofar as the moderation of epistemological pretensions seems to be precisely the point of Locke's endeavour, in Drafts A, B, and C and in the finished *Essay*. Here Locke twisted his usual point rather hard: it is epistemological pretension to declare that we *must* restrict our knowledge to what is proportional to our powers. No such reasoning is applied to the mystery of the incorporeal soul.

The same pattern of separation between the discussion of the constructed idea of God and the presentation of arguments for the existence of God appears in the important chapter II. xxiii of the *Essay*, where Locke first proposed that the 'idea' of God is arrived at by a Gassendi-type process of mental concretion of qualities:

> [I]f we examine the *Idea* we have of the incomprehensible supreme Being, we shall find, that we come by it the same way; and that the complex *Ideas* we have both of God and separate Spirits are made of the simple *Ideas* we receive from *Reflection*; v.g. having, from what we experiment in our selves, got the *Ideas* of Existence and Duration; of Knowledge and Power; of Pleasure and Happiness; and of several other Qualities and Powers, which it is better to have than to be without; when we would frame an *Idea* the most suitable we can to the supreme Being, we enlarge every one of these with our *Idea* of Infinity; and so putting them together, make our complex *Idea of God*. For that the mind has such a power of enlarging some of its *Ideas*, received from Sensation and Reflection, has been already shewed. (E II. xxiii. 33)

But the fact of the existence of God is derived by what Locke claimed to be a logically inescapable inference. God is the necessary condition of there being thought and knowledge. Something eternal must have produced what there is, and this something eternal cannot be matter, or 'incogitative matter', insofar as incogitative matter cannot produce thought. 'I appeal to every one's own thoughts, whether he cannot as easily conceive matter produced by nothing, as thought to be produced by pure matter, when before there was no such thing as thought, or an intelligent being existing?' (E IV. x. 10.)

This move enabled Locke, following Gassendi, to reject the Cartesian incorporeal soul.[5] Thought and knowledge would be impossible in a purely corporeal universe; but, given God, who can superadd the power of thought to organized matter, everything but God may be purely corporeal, an inference Locke appeared to welcome though never explicitly to endorse.

Proof of the existence of this God—a minimal God—has no intrinsic moral importance, at least not any importance that can be demonstrated so easily by philosophy. The God who punishes and rewards is an idea, an idea constructed, a Gassendist might argue, by extrapolating from our worldly experience of punishment and reward and by supposing a being able to carry out what escapes the power and reach of human beings. Yet Locke—because his naturalism acknowledged the power of human desires and the very limited human ability to foresee the future and to care about it (E II. xxi. 34–5) needed to recognize a God with these powers. Also, to escape the moral relativism that would seem to follow from his doctrine of experience, he needed a legislator. Descartes did not feel himself to be under any such pressure.

3

For Descartes, as noted, God scarcely qualified as a person, let alone as a judge or a punitive enforcer.[6] Descartes seemed to endorse the notion that God lays down the law, not only to corpuscles but to human beings as well, at the start of Part 2 of the *Discourse* when he declared that 'peoples who have grown gradually from a half-savage to a civilized state, and have made their laws only insofar as they were forced to by the inconvenience of crimes and quarrels, could not be so well governed as those who from the beginning of their society have observed the basic laws laid down by some wise law-giver. Similarly, it is quite certain that the constitution of the true religion, whose articles have been made by God alone, must be incomparably better ordered than all the others' (AT 6, pp. 12–13; CSM 1, pp. 116–17). However, in its context, the passage seems to be a plea for the familiar Cartesian intellectual autocracy rather than for obedience to the one true religion.

In any case, Descartes's ethics are conspicuously cut off from divine command. He is pragmatic, thus also a relativist, and, like Locke in this regard, a proponent of religious toleration.[7] His moral ideal appears to conform to Pierre Charron's ideal of 'full and universal freedom of the mind', described by E. H. Rice as a feature of late Renaissance scepticism.

[5] Not everyone agrees, but see Wilson, 'Managing Expectations', and Jolley, *Locke's Touchy Subjects*, esp. pp. 67–83.

[6] God's intelligence, power, and veracity lent themselves to a naturalistic interpretation as Spinoza realized; see the comments by Leibniz in Section 5.

[7] Gillespie, 'Descartes and the Question of Toleration'.

This freedom consists in a liberty to judge all things, to be bound by none of them, and to remain universally receptive to every point of view. It is a purely internal and 'secret' freedom which does not involve one's external behaviour in the world. On the contrary, one should conform details of dress, behaviour, and usage to the laws, customs, and ceremonies of the country in which one lives.[8]

The 'provisional morality' of the *Discourse*—outward conformity in inessentials to ensure complete psychological freedom where it matters—is a direct evocation of Charron. Recall Descartes's first resolution:

> ...to obey the laws and customs of my country, holding constantly to the religion in which by God's grace I had been instructed from my childhood, and governing myself in all other matters according to the most moderate and least extreme opinions—the opinions commonly accepted in practice by the most sensible of those with whom I should have to live.
>
> (AT 6, pp. 23–4; CSM 1, p. 122)

Positive ethics is simply control of excessive passion for one's own good and that of others, not the aspiration towards Christian ideals of sinlessness. The aim, one might say, is to be able to proceed through life and to conduct one's studies without being interfered with and without making unnecessary difficulties for oneself.

Locke's theory of morals,[9] by contrast with Descartes's, reflects his view of God as a person with moral concerns, as the very foundation of human morality. His conviction that there were no innate, universal practical principles left Locke out on a limb, threatening the conclusion Descartes by and large accepted: morality is whatever your society says is moral and rewards you rather than punishing you for. Pagan morality, Locke was persuaded, was deeply inadequate: the Greeks and Romans, knowledgeable and sophisticated in so many respects, were people who exposed their children (*Reasonableness*, 196–8; cf. E I. iii. 9). Locke knew of no other religion besides Christianity (perhaps he was right) that prescribes tenderness towards children and some modicum of respect for women, at least as mothers. It was these virtues that he noticed were regularly disregarded, especially by the 'savages' who he thought took a distinctly utilitarian perspective on sex and food.

As Richard Aaron remarks, however, 'It is in vain that we search in [Locke's] pages for a consistent ethical theory'[10]—relating moral law to natural law, to divine will and to our motives. And as Ashcraft emphasizes, Locke's convictions about the one true morality and the assured destiny of those who obey and who do not obey is at odds with his constant emphasis on human ignorance. Locke tried at one point in the *Essay* precisely to differentiate moral knowledge, which could be secured, from our knowledge of substances, which could not, arguing that because moral notions are mixed modes and made by the mind, we can grasp their entailment relations (E III. xi. 17; IV. iii. 18). This attempt was abortive and his fallback position was that morality is what the New

[8] Rice, *The Renaissance Idea of Wisdom*, 188.

[9] For overviews see Sheridan, 'Locke's Moral Philosophy', earlier Colman, *John Locke's Moral Philosophy*.

[10] Aaron, *John Locke*, 266–7.

Testament enjoins. Locke could escape from the incorporeal soul and innate ideas, but he had no choice but to attend to the specifically Christian revelation and its foundation in testimony regarding miracles. Thus faith, a faculty unrelated to the experience and reflection supposed in the *Essay* to be the basis of all knowledge, had to enter the epistemological picture. At the same time, Locke's meticulous attention to scriptural issues and his appeals to faith can obscure the point that Christianity supplies for him only a minimal morality and that it is the force of the *idea* of God, not the actions of God in another world, that actually sustains or ought to sustain human morality.

4

In another respect as well, there is more consistency in Locke than one might at first suppose. The two realms of theoretical and practical knowledge are analogous. As well as being essentially baffled by nature, unable to penetrate to real essences, unable to understand the generation of qualities, helpless in the face of the complexity and interdependence of things, we are morally ignorant and impaired.

> 'Tis not requisite on this occasion, for the inhancing of this benefit, to enlarge on the frailty of our Minds, and weakness of our Constitutions; How liable to mistakes, how apt to go astray, and how easily to be turned out of the paths of Virtue. If any one needs go beyond himself, and the testimony of his own Conscience in this point; If he feels not his own errors and passions always tempting, and often prevailing, against the strict Rules of his Duty; He need but look abroad into any Age of the World to be convinced. (*Reasonableness*, 290)

In the presentation of the *Reasonableness of Christianity* as well as in the *Essay*, human beings are irremediably 'mediocre'—emotional, incompletely rational, partially blind—and they face a confusing world of hidden causes and hidden effects. There are only a few known laws, in physics and in morality, and they don't explain or cover much. Only love to God and charity to ourselves (an interesting notion) are absolutely necessary moral requirements (LC 1, p. 556). When Denis Grenville, an overly scrupulous, morally obsessive man, wrote to Locke in 1678 for assurance that his moral strivings were sufficient, Locke replied, expressing a view of moral knowledge that is parallel to rather than opposed to the main themes of the later *Essay*. We cannot see into the essences of our actions or predict their remote consequences; nor can we imaginatively encompass all the alternatives to them.

> Our short sight not penetrating far enough into any action to be able to discover all that is comparatively good or bad in it: Besides that the extent of our thoughts is not able to reach all those actions which at any one time we are capeable of doing...we cannot secure ourselves from being in the wrong. (LC 1, p. 557)

As Locke later described God as having adapted our sense organs to our ordinary mode of life, though not to the underlying structure of things, here he suggested that God has adapted our desires to his expectations. 'I cannot imagine how God who has compassion on our weaknesse and knows how we are made, would put pore man nay the best of men, those that seeke him with sincerity and truth under almost an absolute necessity of sinning perpetually against him' (LC 1, p. 557). He told Grenville,

> I confesse our duty is sometimes soe evident, and the rule and circumstances soe determin it to the present performance, that there is no latitude left.... But I think this seldome happens. At least I may confidently say it does not in the greatest part of the actions of our lives wherein I think god out of his infinite goodnesse considering our ignorance and frailty hath left us a great liberty. (LC 1, p. 557)

The contrast between human justice, which works with crude criteria of identity and responsibility, and divine justice which employs finer ones, seems to assure the reader that God will punish people in the hereafter for actions that escaped punishment in the world, including forgotten ones and those performed in a state of absent-mindedness. Yet Locke's aim was originally—before the issue of 'culpable drunkeness' was raised by Molyneux—to establish that 'no one shall be made to answer for what he knows nothing of' (E II. xxvii. 22).

The merciful disposition of God is a point reinforced in Locke's first Letter on Toleration and in the *Reasonableness* in which he asserts that Christ

> did not expect...a Perfect Obedience void of all slips and falls: He knew our Make, and the weakness of our Constitutions too well, and was sent with a Supply for that Defect. Besides, perfect Obedience was the Righteousness of the Law of Works; and then the Reward would be of Debt, and not of Grace; And to such there was no need of Faith to be imputed to them for Righteousness. (*Reasonableness*, 213)

Such views were anticipated and shared by the softer wing of Anglican theology and by the Puritan Richard Baxter, as well as by the 'Latitude-Men'. Henry Hammond's *A paraphrase and annotations upon all the books of the New Testament briefly explaining all the difficult places thereof* of 1659, the first full-length commentary, anticipated Locke's doctrine of the proportionality of our perceptions and moral capabilities to the life-world and of the contingency of perfection as depending on grace and inachievable by man on his own.[11] Even Edward Stillingfleet, who strenuously objected to all traces of Epicurean corporealism and hedonism in Locke, admitted moral fallibilism.[12]

[11] '[T]he Illuminations ordinarily afforded by God are proportioned not to his all-seeing knowledge, but to our capacities, and our real wants, and so, as his Sanctifying grace is not given in such a degree, or manner, as to preserve us impeccable, so neither his Illuminations, as to render us inerrable or infallible.' So, Hammond concluded, 'where God affords not his grace, he requires not of us those performances to which that grace was necessary, and so...he will supply by his pardon what was wanting in our strength...and...will pardon errors of weakness as well as sins of weakness, humane nesciencies, as wel as humane frailties...' (*A paraphrase*, 10).

[12] In discussing exaggerated righteousness, Stillingfleet declared it an error not to 'Make allowance for the common Infirmities of Mankind; which do not only consist in the Imperfections of good Actions, but in such Failings, which human Nature is subject to in this State; notwithstanding our greatest Care to avoid them.... We must in judging others make allowance for the Weakness of Judgment and Strength of Passion, which we find often accompanied with a real Tenderness of Conscience.... The best use we can make of this, is, to pity the Infirmities of Mankind, and of those most, who are under the Conduct of

5

A corresponding notion of 'Cartesian humility' might seem difficult to coax from Descartes's texts. Descartes emphasized rather the epistemic qualifications he possessed through having been made in God's image (AT 7, p. 51; CSM 2, p. 35). The power of the human will is said by the Meditator to be 'so great that the idea of any greater faculty is beyond my grasp; so much so that it is above all in virtue of the will that I understand myself to bear in some way the image and likeness of God' (AT 7, p. 58; CSM 2, p. 40). True, Descartes admits that memory, imagination, and understanding are all 'weak and limited', and the at times overweening pretension of the *Principles* is qualified by the point that 'our mental capacity is very mediocre' (AT 9A, p. 80; CSM 1, p. 248). But the point he is trying to make in this connection is that we cannot know that 'God made everything for our benefit' (AT 9A, pp. 80–1; 248) and that readers should not dismiss the Cartesian account of the origins of the world because they think they know otherwise about the Creation. As in Locke's flotation of the notion of 'thinking matter' (E IV. iii. 6) nescience can be a Trojan horse.

In Cartesian ethics, the emphasis is not on the frailty of human nature, the weakness of the will, and the responses to it of a just but forgiving God, but rather on God's utter implacability. In the 'Conversation with Burman' (1648), Descartes referred to God's 'indifference', and to the futility of prayer: 'God is... quite unalterable, and... has agreed from eternity either to grant me a particular request or not to grant it...' (AT 5, p. 167; CSMK, p. 348). He went further in suggesting that we cannot either venerate or scorn God, but only love him. 'Veneration or respect', he pointed out, 'is an inclination of the soul not only to have esteem for the object that it reveres but also to submit to it with some fear in order to try to gain its favour.'

> Accordingly we have veneration only for free causes which we judge capable of doing us good or evil, without our knowing which they will do. For we have love and devotion rather than simple veneration for those causes from which we expect only good, and we have hatred for those from which we expect only evil. And if we do not judge the cause of this good or evil to be free, we do not submit to it in order to try to gain its favour. Thus, when the pagans had veneration for woods, springs, or mountains, it was not strictly speaking these dead things that they revered, but the divinities which they believed to preside over them.
>
> (AT 11, pp. 454–5; CSM 1, p. 388)

We cannot venerate God, nor can we fear God, because God is not a free cause. To consider God as capable of doing us harm or favours, as interested to gratify or chastise us, is to think as the pagans do.[13] But we can love God, and we can, Descartes thinks,

a mis-guided Conscience, because whatever the Action be their Design and Intention is honest and good.' *Thirteen sermons*, 513.

[13] The characterization of God as other than a 'free cause' might raise a question as to its consistency with Descartes's alleged extreme voluntarism, allegedly manifested in his claim that God is the author of the eternal truths (AT 1, p. 149). But it is also the case that for Descartes, once God has ordained something, he does not waver, and all events flow from God's initial act of will. The fatalistic tenor of his philosophy

accept everything that happens to us in a good frame of mind. We can derive joy 'from everything' when we realize that sorrows as well as joys, pains as well as pleasures, come from God, that is to say from the infinite power and rationality that God/Nature is, from seeing ourselves as part of this system, its laws working on us and through us. The doctrine of which Descartes was so proud—that God is the author of all the laws, moral, physical, and logical, including the eternal truths—is a statement of their inexorability. Here Descartes echoed the Stoic Justus Lipsius, an author well known to him:

We must acknowledge him to be stayed, resolute and immutable, alwaies one, and like himselfe, not wauering or varying in those thinges which once he willed and foresawe. For, The eternall God neuer chaungeth his minde, saith Homer. Which if thou con|fesse to bee true (as needes thou must, if there be in thee any reason or sense) this also must be allowed that all Gods decrees are firme and immooueable euen from euerlasting vnto all eternitie: of this groweth necessitie, and that same DESTINIE which thou deridest.[14]

We cannot see the whole or fully appreciate the 'vastness, beauty, and perfection' of the universe (AT 9B, p. 80; CSM 1, p. 248). In a letter to Chanut of 1647, Descartes retreated from the scientific and metaphysical presumptions of the *Meditations* and the *Principles* to an ethical modesty:

We must weigh our smallness against the greatness of the created universe, observing how all created things depend on God, and regarding them in a manner proper to his omnipotence... If a man meditates on these things and understands them properly, he is filled with extreme joy. Far from being so injurious and ungrateful to God as to want to take his place, he thinks that the knowledge with which God has honoured him is enough to make his life worthwhile.

(AT 4, p. 609; CSMK, p. 309)

The wise man,

because he know that nothing can befall him which God has not decreed,...no longer fears death, pain, or disgrace. He so loves this divine decree, deems it as just and so necessary, and knows that he must be so completely subject to it that even when he expects it to bring death or some other evil, he would not will to change it even if—per impossibile—, he could do so. He does not shun evils and afflictions, because they come to him from divine providence; still less does he eschew the permissible goods or pleasures he may enjoy in life, since they too come from God. He accepts them with joy, without any fear of evils, and his love makes him perfectly happy. (AT 4, p. 610; CSMK, p. 310)

was not only identified by Leibniz, and carried further by Spinoza, it is backed up by certain remarks in the Correspondence with Elisabeth, where Descartes claims that even his own thoughts are foreseen and desired by God (AT 4, p. 314).

[14] Lipsius, *Of Constancie*, 42. This is a form of 'soft' Stoicism by contrast with the rather 'hard' Stoicism of Epictetus. It recalls the aestheticism and detachment of Marcus Aurelius, a figure barely known before 1650. The lawfulness of nature extends to the human realm and has its own beauty: For Marcus, 'Everything that happens is as customary and understandable as the rose in springtime or the fruit in summer. The same is true of disease, death, slander, and conspiracy, and all the things which delight or pain foolish men' (*Meditations*, 34).

We grasp that all changes and motions come from God, the source of the eternal truths upon which everything else depends. So it is with events as Hippocrates said of diseases—all are natural, all are divine.[15] Disease, for Descartes, is simply an element of the divinely established order. 'Even when we are ill, nature still remains the same' (AT 5, p. 179; CSMK, p. 354); 'a sick man is no less one of God's creatures than a healthy one, and it seems no less a contradiction to suppose that he has received from God a nature which deceives him' (AT 7, p. 84; CSM 2, p. 58).

Leibniz purported to be exasperated by the combination in Descartes of determinism and voluntarism. Determinism for Leibniz was morally and intellectually acceptable only if one could have ethical confidence in the world, in which case its goodness had to consist in more than God's decision to create it as he did. In a Letter to Molanus of *c.*1679, he complained that 'A God like Descartes's allows us no consolation other than that of patience through strength . . . [E]ach creature will be happy or unhappy depending on how it finds itself engulfed in these great currents or vortices. Descartes has good reason to recommend, instead of felicity, patience without hope' (Gerhardt, ed. 4, p. 297–303; Ariew and Garber, trans. 242).

6. Conclusion

For all the evident differences between them, not only in the epistemological sphere, but in their theologies and moral stances, the 'essential religiosity', as distinct from formal theology, of both Descartes and Locke is associated with the notions of forgiveness and non-judgementalism. The authoritarianism, censoriousness, and scrupulosity of the old-style theologian are absent from their writings. There is enhanced psychological and anthropological realism, especially in their treatments of the passions.

Descartes's attitude of acceptance towards all that befalls us followed from his conception of the universe as a law-governed realm, manifesting God's impersonal wisdom and power. Contrary to the accusations of Leibniz, and the extrapolations made by Spinoza, the world for Descartes is good as opposed to ethically neutral. The human emotional system, like the human perceptual system, is adapted to the world. This adaptation can be seen as a provision of nature, perhaps the effect of the long history of the universe and the succession of conditions realized under the operations of the laws of nature. Or it can be seen as a divine provision, reflecting God's goodness and concern for living and thinking beings. In either case, we are healthiest when we live according to nature, that is when we 'live and eat like the animals . . . i.e. as much as we enjoy and relish, but no more' (AT 5, p. 178; CSMK, p. 353).

A pathos is defined by the Stoic Zeno as 'an agitation of the soul alien from right reason and contrary to nature'.[16] This is entirely wrong, according to Descartes. Because they failed to understand the physiological basis, usefulness, and pleasurable quality of

[15] In connection with epilepsy, the 'divine disease'. *Hippocrates*, 2.139–83.

[16] Cicero, *Tusculan Disputations* 4. 6. 11.

the passions, the Stoics propounded untenable views about *apatheia* and self-control (AT 11, p. 327; CSM 1, p. 328). An emotion is simply a response to a surprise, and, as mind–body unities we are mostly unprepared for the 'contingencies of life'. '[T]he objects of the passions produce movements in the blood which follow so rapidly from the mere impressions formed in the brain and the disposition of the organs, without any help at all from the soul, that no amount of human wisdom is capable of counteracting these movements when we are not adequately prepared to do so' (AT 11, p. 486; CSM 1, p. 403). The conclusion to the *Passions of the Soul* is that one ought to manage the passions with such skill that 'the evils they cause are quite bearable and even become a source of joy' (AT 11, p. 488; CSM 1, p. 404). The quasi-divine powers of the human will cannot be recruited to this end; they are restricted by Descartes to intramental operations, especially drawing inferences.

Locke, too, emphasizes, rather daringly, the power of our natural desires. The will is never moved to action except by 'uneasiness' (E II. xxi. 31–40). For Locke, 'We are liable to forms of "extreme disturbance"' that 'possess our whole Mind, as when the pain of the Rack, an impetuous *uneasiness*, as of Love, Anger, or any other violent Passion, running away with us, allows us not to liberty of thought, and we are not Masters enough of our own Minds to consider thoroughly, and examine fairly' (E II. xxi. 53). His belief that God is merciful with respect to human weakness and our tendency to stumble and blunder follows from his sense of the complexity of nature and human affairs and the mediocrity of human reason. Although his advocacy for religious toleration does not extend as far as Descartes's and indeed excludes Descartes's view of morality as convention, his anthropological realism and his attempt to lighten the sense of sin associated with the passions is as impressive as Descartes's.

The moral shift in the following century, from a concern with the depravity of individuals and the distortion of their proper relationship to God to a concern with the corruption of social institutions and practices and the distortion of proper relationship between human and human, is a complex story. Both Descartes's presentation of the human being as caught up rather helplessly in nature, but as enjoying the power to understand it and thereby to control it by 'reason', and Locke's presentation of human beings as victims of their desires and their ignorance—but, for all that, as eminently educable, and as makers of their concepts and institutions—had a part to play.

References

Aaron, Richard. *John Locke*. Oxford: Clarendon, 1973.

Ashcraft, Richard. 'Faith and Knowledge in Locke's Philosophy', in John W. Yolton, ed., *John Locke: Problems and Perspectives*. Cambridge: Cambridge University Press, 1969: 194–223.

Ayers, Michael. *Locke*, 2 vols. London: Routledge, 1993.

Cicero, Marcus Tullius. *Tusculan Disputations*, trans. J. E. King. Cambridge, Mass.: Harvard University Press, 1927.

Colman, John. *John Locke's Moral Philosophy*. Edinburgh: Edinburgh University Press, 1983.

Gillespie, Michael Allen. 'Descartes and the Question of Toleration', in Alan Levine, ed., *Early Modern Skepticism and the Origins of Toleration*. Lanham, Md.: Lexington Books, 1999: 103–20.

Hammond, Henry. *A paraphrase and annotations upon all the books of the New Testament briefly explaining all the difficult places thereof*. London: J. Flesher, 1659.

Hippocrates. *Hippocrates*, trans. W. H. J. Jones, 2 vols. Cambridge, Mass.: Harvard University Press, 1959.

James, Susan. *Passion and Action: The Emotions in Seventeenth-Century Philosophy*. Oxford: Clarendon, 1997.

Janiak, Andrew. 'Metaphysics and Natural Philosophy in Newton and Descartes', *Foundations of Science*, 18 (2013), 403–17.

Jolley, Nicholas. *Locke's Touchy Subjects: Materialism and Immortality*. Oxford: Oxford University Press, 2015.

Leibniz, G. W. *Philosophische Schriften von Leibniz*, ed C. I. Gerhardt, 7 vols. Hildesheim: Olms, 1965.

Leibniz, G. W. *Philosophical Essays*., trans. and ed. Roger Ariew and Daniel Garber. Indianapolis: Hackett, 1989.

Lipsius, Justus. *Of Constancie (De Constantia)*, trans. John Stradling. London, 1595; repr. Bristol: Bristol Phoenix Press, 2006.

Marcus Aurelius. *The Meditations*, trans. G. M. A. Grube. Indianapolis: Hackett, 1983.

Nadler, Steven M. 'Arnauld, Descartes, and Transubstantiation: Reconciling Cartesian Metaphysics and Real Presence', *Journal of the History of Ideas*, 49/2 (1988), 229–46.

Rice, E. H. *The Renaissance Idea of Wisdom*. Cambridge, Mass.: Harvard University Press, 1958.

Sailor, Danton B. 'Cudworth and Descartes', *Journal of the History of Ideas*, 23 (1962), 133–40.

Shapin, Steven. *A Social History of Truth*. Chicago: University of Chicago, 1994.

Sheridan, Patricia. 'Locke's Moral Philosophy', in Edward N. Zalta, ed., *The Stanford Encyclopedia of Philosophy*, <http://plato.stanford.edu/archives/sum2014/entries/locke-moral/>.

Stillingfleet, Edward. *Thirteen sermons preached on several occasions three of which never before printed/by the Right Reverend Father in God Edward, Lord Bishop of Worcester*. Henry Mortlock: London, 1698.

Wilson, Catherine. 'Locke's Moral Epistemology', in L. Newman, ed., *The Cambridge Companion to Locke's* Essay. Cambridge: Cambridge University Press, 2007: 381–405.

Wilson, Catherine. 'Managing Expectations: Locke on the Material Mind and Moral Mediocrity', in Anthony O'Hear, ed., *Royal Institute of Philosophy Supplements*. Cambridge: Cambridge University Press, 2016.

10

Locke and Port-Royal on Affirmation, Negation, and Other Postures of the Mind

Laurent Jaffro

Is it possible to reconcile these two apparently divergent facts?

(1) Locke attributes to affirmation and negation the task of marking the connection or 'separation' of ideas that constitutes the proposition and at the same time the task of marking the assertion by which we express our assent. For this reason, he would find himself in great difficulty, given that if each proposition supposes this affirmation or negation, then each proposition would be necessarily asserted as soon as it was formed, which would leave no room for other attitudes such as doubt, as Michael Ayers has noticed.[1] According to David Owen, who ascribes to him the view that 'there is only one act of the mind' involved in predication and in affirmation, it is a price that Locke has to pay for that view.[2]

(2) Locke allows however a great diversity of ways in which the mind can consider a proposition. This diversity is, for him, particularly shown by the many uses and meanings of syncategorematic terms, 'particles', with which affirmation and negation have strong affinity (E IV. i. 1).

In short, how can we combine in a coherent manner that which Locke's epistemology tells about affirmation and negation with that which Locke's philosophy of language tells, in a very allusive manner, about syncategorematic terms?

Walter Ott has drawn attention to the importance of the question of particles in Locke's philosophy of language. In the chapter he dedicates to 'Particles and propositions', he deals with particles in the first section, whilst in the second he attempts to defend, against Owen and Ayers, the thesis of the distinction between assertion and

[1] Ayers, *Locke. Volume I: Epistemology*, pp. 104–5.
[2] Owen, 'Locke and Hume on Belief, Judgment and Assent', pp. 17–18.

predication as two acts of the mind.[3] However, in the second section, to support the two-act thesis, Ott tends to rely on considerations linked to psychology and epistemology, rather than on the theory of language. My hypothesis, on the contrary, is that it is necessary to look at the philosophy of language to understand how what Locke says concerning affirmation and negation does not exclude that they signify a diversity of actions of the mind.

The comparison with Port-Royal (meaning Arnauld and Lancelot's *Grammaire générale et raisonnée* and Arnauld and Nicole's *La Logique ou L'Art de penser*) is useful because it is clear that Port-Royal and Locke are in the same tradition with regard to their logical and grammatical thoughts on the syncategorematic terms, which signify operations on conceptions and not the conceptions themselves.[4]

My intention is to establish:

(a) That, if it is a difficulty, induced by the conception of judgement (in the Cartesian sense of *La Logique ou L'Art de penser*, which applies as much to that which Locke calls 'knowledge' as to that which he calls 'judgement') as a connection of ideas, to not be able to distinguish assertion as a propositional attitude from predication as a subpropositional operation, and to confuse under the names of affirmation and negation these two operations, this difficulty is as much a problem for Port-Royal as it is for Locke (or, if one prefers, as small a problem for Locke as it is for Port-Royal);

(b) That this conception of judgement does not prevent Port-Royal and Locke from presenting (Port-Royal) or outlining (Locke) a theory of language in which there is room for affirmation (and negation) to denote distinct acts of the mind.

1. Affirmation and Negation as Markers of Predication

To begin, we should pay attention to the fact that the one-act thesis is very plausible within the perimeter of intuitive knowledge. My main contention will be that it does not apply in other areas of Locke's philosophy of assent. Leaving intuitive knowledge of one's existence aside, intuitive knowledge is the perception without mediation either (m) of an identity or difference between ideas, or (n) of a 'relation.' Let us consider case (m): 'When we know that *White is not Black*, what do we else but perceive, that these two *Ideas* do not agree?' (E IV. i. 2; see also E IV. ii. 1). The mind intuitively perceives, without intermediary proof, through only the consideration of the ideas of white and black, that white is not black. Thus the mind simultaneously asserts that white is not

[3] Ott, *Locke's Philosophy of Language*, pp. 38–40, 47–9.

[4] On the distinction between objects of thought and manners of thinking in Port-Royal's *Grammar*, and on the fact that 'affirmation' proper does not stand for an idea but for an act of affirmation, see Nuchelmans, *Judgement and Proposition*, pp. 73–4; on Locke's similar views, see ibid. p. 146.

black and predicates of white that is not black. The absence of gap between the two is typical of intuitive knowledge.

To *perceive* that white is not black, is to think and possibly say something about white, that it is not black, and also to know, and thus assert, that white is not black. Locke describes both operations as one act of separating the idea of white and the idea of black. For Locke's proposal to be plausible, the 'separating' must be more than a non-'joining'; it is an opposite operation and not simply the absence of junction (if not, each time that we were not in the process of joining the idea A to the idea B, we would be in the process of denying that B was A). And the junction must be more than the non-separation (if not each time that we were in the process of having simultaneously the ideas A and B, we would be in the process of affirming the one of the other). The reason why predication and assertion are indistinguishable in this case of intuitive knowledge resides in the nature of the agreement or disagreement between ideas, that is, identity or difference, which immediately determines affirmation or negation. The difference between *x* and *y* is reflected by the mental or verbal separation of *x* and *y*, and brings about the (assertive) perception that *x* is not *y*. Not only 'separation' bears both functions, but the distinction between the two is psychologically indiscernible.

Now, we have precious information on the diversity of the foundations of linguistic or mental operations of affirmation and negation. In E IV. i. 4, Locke lists four sorts of agreement or disagreement between our ideas: identity or difference (of which an example is '*White is not Black*'); relation; coexistence or necessary connection; real existence. He interestingly indicates that if the only type of agreement or disagreement between our ideas were identity or difference, in the sense that each idea is identical to itself and different from all other ideas (numerically), then all that we could formulate concerning distinct ideas would be negative propositions: we would only be saying about an idea that it was not another idea and our knowledge would be reduced to the repetition of mental acts that logicians summarize in the principle of identity. Happily though, there are other 'sorts' of agreement and disagreement. We thus learn something indirectly about the operations of joining and of separating: leaving aside propositions that spell out identity, when this operation is correctly carried out, that is, when the predication constructs a proposition which is true, then it reflects either a relation or a rupture of relation, or a coexistence of properties in a substance or the fact of non-coexistence. These different types of agreement or disagreement are what Locke calls the 'different grounds of Affirmation and Negation' in E IV. i. 7. For this formula to have a sense, it is necessary that the operations of affirmation and of negation be distinct from the apprehension of the agreement or disagreement that makes the 'grounds'.

Let us now take an example from outside the realm of intuitive knowledge. The proposition '*Iron is susceptible of magnetic Impressions*' (E IV. i. 7) is constructed by the junction between the subject and the predicate. But in this case the junction is founded, justified, by our experience of the coexistence of the properties mentioned in this metallic substance. It is because it is thus justified for us that we take this proposition to be true. In this case there cannot be truth or falsity without an affirmation or a negation, primarily because affirmation and negation are the formal conditions of truth or falsity: according to the theory of judgement as an operation on ideas that affirms or denies one term or another, these formal conditions are not fulfilled at the level of ideas but at the level of judgement. For Locke, a conception cannot be true or false unless it says something, that is to say unless it is a hidden judgement, formally made up of an affirmation or a negation. In addition, this proposition is true when the affirmation or negation that constitutes it reflects the agreement or disagreement between the things signified. In the case of our judgements on the properties of individual substances, affirmation or negation has an epistemic function: if they are well formed, they translate the structure of reality into the structure of the proposition.

To understand the diversity of the occurrences of affirmation and negation, it must be remembered that affirmation and negation also play an important role in the formation of our complex ideas of substances, which can be false in two different ways: when they combine ideas of quality which are not combined in nature or when they separate 'by a direct Negation' ideas of qualities which are combined in the substance of which we speak (E II. xxxii. 18). Affirmation and negation are thus operations that are components of complex ideas of substances. I can have a complex idea of a substance, which rests on such an affirmation or negation, without however asserting the implicit proposition that this idea contains: I can for example decide to test this implicit proposition by conducting chemical experiments.

We should have this in mind when reading this passage:

> Though Truth and Falsehood belong, in Propriety of Speech, only to Propositions; yet *Ideas* are oftentimes termed *true* or *false* (as what Words are there, that are not used with great Latitude, and with some deviation from their strict and proper Significations?) Though, I think, that when *Ideas* themselves are termed true or false, there is still some secret or tacit Proposition, which is the Foundation of that Denomination: as we shall see, if we examine the particular Occasions, wherein they come to be called true or false. In all which, we shall find some kind of Affirmation, or Negation, which is the Reason of that Denomination. (E II. xxxii. 1)

Here Locke has in mind various sorts of ideas, among which the complex ideas of substance:

> *Ideas* of Substances are, in this respect, also *false*, when, from any Collection of simple *Ideas*, that do always exist together, there is separated, by a direct Negation, any other simple *Idea*, which is constantly joined with them. (E II. xxxii. 18)

In this case, because having a complex idea of a substance consists in thinking of some x as being m, n, $o \ldots$ and/or not being u, v, $w \ldots$, it is clear that each time we consider such an idea, for instance the idea of a piece of metal as being stuff that is (or is not) susceptible of magnetic impressions, it contains an affirmation or a negation, which constitutes a 'secret or tacit proposition'. I believe that the need that we have of complex ideas of substance, which we revise according to our experience, provides an example of a situation where it is clear that affirmation and negation mark a predication without also automatically marking an assertion.[5]

To sum up: Most often, affirmation (or negation) combines that which we may call, at least provisionally, 'functions': that of predication, that of assertion. It is also clear that in the case of intuitive knowledge, the immediacy of perception is such that this combination takes the form of an apparent identity. A first question was whether sometimes affirmation and negation do not combine both functions. A second question is whether the combination of functions is evidence of a faulty philosophy (a confusion, Locke not having had the chance to learn from Frege, Russell, or Geach about the distinction between propositional attitude and propositional content), or, rather, whether it depends on a normal capacity of certain terms, which early modern theories of language can account for.

To the first question, the answer was positive, upon two considerations: (1) Locke's use of the notion of 'grounds of Affirmation and Negation', which is intelligible only on the condition that the justifying level (the 'grounds') of the comparison of ideas is distinguished from the level of affirmation and negation which are then the markers of an assertion thus justified. (2) The use of affirmation and negation as markers of predication in the complex ideas of substance, which can be considered and tested, thus presupposing that affirmation and negation do not necessarily mark assertion.

I add a third, textual, consideration. There exists at least one passage, in Locke's *Second Vindication of The Reasonableness of Christianity*, in which it is explicit that assertion is not confused with affirmation. The context is religious persuasion. Some people would like us to swallow verbal theological propositions the terms of which are loaded, by them, with mysterious significations that *we* cannot see:

> The Question is about a Proposition to be believed, which must first necessarily be understood. For a Man cannot possibly give his assent to any Affirmation or Negation, unless he understand the terms as they are joyn'd in that Proposition, and has a Conception of the thing affirm'd or deny'd, and also a Conception of the thing, concerning which it is affirm'd or deny'd, as they are put there together. But let the Proposition be what it will, there is no more to be understood, than is expressed in the terms of that Proposition.[6]

To speak thus of an 'assent' given to 'affirmation or negation' evidently presupposes that the act of assent is here completely distinct from the operation of affirmation or of negation. Besides, the understanding that is required as a precondition of assent, as much on the level of conceptions ('as they are put together') as on the level of terms ('as they are joined in that proposition'), is an understanding not of each term or conception signified alone, but of their connection. If you say to me 'this is the body of Christ', it is necessary (a condition necessary but not sufficient of my assent or denial

[5] This may be the reason why Locke uses the qualification '*some kind of* Affirmation, or Negation' (E II. xxxii. 1).

[6] Locke, *Vindications of The Reasonableness of Christianity*, p. 80.

or of my suspension) that I understand not only each of the terms, but also their connection. If not, my understanding of that which you say to me would be the same as if you had said 'this is not the body of Christ'.

2. Affirmation and Negation as Markers of Assertion

It is now time to turn to the second question about the status of assertion. Let us get back to the example taken from the first 'sort' of agreement and disagreement that may be the object of intuitive knowledge: identity or diversity in ideas (E IV. i. 4). To perceive that white is not black is, among other things, to assert the truth of the proposition *white is not black*, or to assert the falsity of the proposition *white is black*;[7] but Locke interprets this assertion as a negation that white may be said of black. That is the reason why he does not verbally distinguish assertion from affirmation, nor does he distinguish negative assertion (denial) from negation except when he uses the lexicon of assent or of faith to deal with a specific type of assertion, that which is relative to the junction or separation of ideas that are literally 'proposed' to me and that present themselves to me as verbal propositions.[8] This function of assertion is obviously present in affirmative propositions that form the subject of intuitive knowledge and each time they form the subject of this knowledge. In practice, it is impossible to know intuitively that A is B and not to assert that A is B.

It is to be noted that in contemporary philosophy, because of the popularity of the distinction between propositional attitude and propositional content, the standard view is that the function of assertion, in the form of an act of the mind, is external to its propositional object. It is beyond the scope of this chapter to determine to what extent (obviously weak), what Locke calls 'proposition' is the same as what, since Frege, has been called proposition. It should be noted that according to Arnauld, in his contribution to the *Grammaire générale et raisonnée* as well as in his contribution to *La Logique ou L'Art de penser*, the function of assertion (to my knowledge, this term is not used by Port-Royal) is often implicitly borne by the verb, itself reducible to the copula.[9] The manner in which assertion is borne by the verb certainly depends upon the context of enunciation as much as upon the structure of the connections between propositions. For example, if you tell someone that 'white is not black', you can say it in a way that leaves no doubt as to the fact that you are asserting it, whereas if you read to someone the present sentence in full, it is clear that you are not asserting it. In spite of the contextual aspect of the way assertion is borne by the verb, it is obvious that, insofar as

[7] See E I. ii. 18: '*White is not Black*' is one of those propositions to which one 'must necessarily assent' 'at first hearing'.

[8] See my 'Pourquoi l'objet de l'assentiment est-il, selon Locke, la proposition verbale?', in Jaffro (ed.), *Croit-on comme on veut?*, 2013.

[9] See *Logic or the Art of Thinking*, II. 3, pp. 82–3: 'If I say, "God is not unjust", the word "is" when joined to the particle "not" signifies the action contrary to affirming, namely denying, in which I view these ideas as repugnant to one another, because the idea "unjust" contains something contrary to what is contained in the idea "God".'

it is borne by a term, assertion is somehow internal to the proposition; the asserting act of the mind puts ideas together either by separating them (negative assertion) or by joining them (affirmative assertion). If the assertion is borne by an internal element of the proposition, then we understand why authors like Arnauld, and also Locke, seem to us to adhere to the strange conception whereby assertion and predication are in fact the same act, which is inbuilt in the proposition. Assertion and predication would consist of the unique act of affirmative or negative connection of ideas.

The one-act interpretation, that is, the thesis that Locke completely merges predication and assertion, does not pay sufficient attention to the diversity of functions that syncategorematic terms fulfil. As will be shown in Section 3, they can serve the connection as well as the assertion, but also in some cases, as we have seen, the connection without assertion.

The scope of the example 'white is not black' is limited because it involves an object of intuitive knowledge. The operation by which I assert that white is not black is indiscernible from that by which I see that white is not black and thus separate one idea from the other. This act of the mind is a response to the obvious and for that reason it is, according to Locke, completely involuntary. It is the separation of ideas that somehow takes the initiative and brings with it my assertion:

> *Knowledge*, as has been shewn, consists in the perception of the agreement or disagreement of *Ideas*: Now where that agreement or disagreement is perceived immediately by it self, without the intervention or help of any other, there our *Knowledge is self-evident*. This will appear to be so to any one, who will but consider any of these Propositions, which, without any proof, he assents to at first sight: for in all of them he will find, that the reason of his Assent, is from that agreement or disagreement, which the Mind, by an immediate comparing them, finds in those *Ideas* answering the Affirmation or Negation in the Proposition. (E IV. vii. 2)

In this last sentence, I do not see that which suggests that the assertion be identical to the construction of the proposition (*pace* David Owen[10]). At the very most, they are phenomenologically indiscernible because of the immediacy of the comparison. It should be noted that the expression *reason of his assent* echoes the expression *grounds of affirmation and negation*, and thus indicates a distinction between the level of comparison of ideas and that of assent or dissent, although in the case of intuitive knowledge there is no temporal lag between the two.

So it is doubtful that we may learn much about Locke's understanding of assent by considering only (or mainly) his account of intuition. Locke's theory of the operations of the mind, which is the issue here, inasmuch as one can reconstruct the elements through allusions of the *Essay*, permits gradations and is strongly contextualist. The analysis of 'white is not black' does not furnish us with Locke's theory on these subjects, but is a case example against which we should compare other cases.

[10] See Owen, 'Locke and Hume on Belief, Judgment and Assent', p. 25: 'One doesn't perceive the agreement, and *then* assent to the truth of the proposition.' Owen infers from that psychological immediacy a structural identity between proposition formation and assent.

This case example, in fact, does not provide the paradigm of assertion for a very simple reason: the assertion that white is not black is completely trivial and teaches us absolutely nothing. We have no more reason to say of white that it is not black, than to say that it is not square, or that it does not smell bad, or that it is not a syllogism:

> For since all distinct *Ideas* must eternally be known not to be the same, and so be universally and constantly denied one of another, there could be no room for any positive knowledge at all, if we could not perceive any relation between our *Ideas*, and find out the Agreement or Disagreement, they have with one another, in several ways the Mind takes of comparing them. (E IV. i. 5)

As well as having himself erected intuitive knowledge as a paradigm (but it is above all the intuitive knowledge that is the perception of an agreement such as 'relation', and which is presupposed by all demonstrative knowledge, rather than by this other, sterile, intuitive knowledge which is the perception of identity or diversity), Locke invites us to go past it in Book IV in showing the importance of that which he calls 'judgement', opinion, belief, assent, in parallel to that which comes from knowledge per se. Locke's theory, which remains very largely implicit, must therefore allow gradations in the relation between predication and assertion, from complete identity in the case of intuitive knowledge, to clear distinction in the case of prudent use of complex ideas of substance; at the other extreme of this gradation, predication is completely detached from assertion: this is what happens in situations where the question is how one should respond to verbal propositions that are submitted to one's examination. If someone says to me, on showing me a piece of bread, that this is the body of Christ,[11] I can compare the ideas of this piece of bread and of the body of Christ, which allows me to understand the proposition (for to understand this proposition supposes more than to have simultaneously in the mind the idea of a piece of bread and the idea of the body of Christ, I should also consider their relationship and the grounds of affirmation and negation), but that does not, by itself, mean to say that I assert it. The problem is that, to make probable this gradualistic interpretation, which does not conform to that which is suggested by the paradigm of intuitive knowledge, it is necessary to show that it is supported by important and explicit arguments in the *Essay*. These arguments relate to the diverse significations of syncategorematic terms.

3. The Equivocity of Syncategorematic Terms

Noticeably affirmation and negation have multiple missions. Not only to mark the constitutive connection of the proposition (without which the formal conditions of truth and falsity are not present); but also to mark the assertion (by which we express that we hold to be true or false that A is B). And to that we should also add, as we have seen, that affirmation and negation are also given an epistemic role, since the truth of a

[11] See E IV. xx. 10 on the doctrine of transubstantiation.

tacit proposition, which a complex idea of substance consists of, materially depends upon the conformity of the affirmation or of the negation, respectively, with the agreement or disagreement between the ideas or qualities. This is a lot of functions for a single term; now one notices that this variety of functions cannot be reduced to one function.

To the extent that the terms of language are the means for the speakers to 'mark',[12] for themselves and for others, their thoughts, it is perhaps more reasonable to abandon this vocabulary of 'function' (as if a term had a function independently from the intention of the speaker) and to speak of the variety of acts of the mind. Let us consider the hypothesis that linguistic affirmation (or negation) consists in a term that signifies different acts of the mind, which in some, but not all, contexts converge. What makes me suggest such a hypothesis is, so to speak, a distortion in the usual parallel between the mental and the verbal, which happens with syncategorematic terms. Although one might think that, since just one term, on the verbal level, corresponds to both asserting (the proposition) and 'joining' (ideas), there must be just one corresponding act on the mental level. But this would be a wrong conclusion. Whilst, to be understood without confusion, the equivocity of categorematic terms should be avoided as much as possible, the equivocity of syncategorematic terms, on the other hand, is actually recommended and even necessary in order to understand the various ways in which we consider a proposition or a group of propositions. Now affirmation or negation, as we are reminded by a historian of scholastic and early modern logic and theory of language, mark not a conception but an operation of the mind, an exercise on conceptions.[13] So, as with all syncategorematic terms (not only intrapropositional, but also interpropositional terms), the same affirmation (or negation) can signify different acts of the mind:

> Besides Words, which are names of *Ideas* in the Mind, there are a great many others that are made use of, to signify the *connexion* that the Mind gives to *Ideas, or Propositions, one with another*. The Mind, in communicating its thought to others, does not only need signs of the *Ideas* it has then before it, but others also, to shew or intimate some particular action of its own, at that time, relating to those *Ideas*. This it does several ways; as, *Is*, and *Is not*, are the general marks of the Mind, affirming or denying. But besides affirmation, or negation, without which there is in Words no Truth or Falshood, the Mind does, in declaring its Sentiments to others, connect, not only the parts of Propositions, but whole Sentences one to another, with their several Relations and Dependencies, to make a coherent Discourse. (E III. vii. 1)

Even though there is an important difference between the marks of affirmation and negation (which, as I understand Locke, 'signify the *connexion* that the Mind gives to Ideas') and particles proper (which 'connect not only the parts of Propositions, but whole Sentences one to another'), affirmation and negation are terms that Locke puts in the same category as particles, according to the logico-grammatical tradition, which

[12] See for instance E II. xi. 9: Words are used as 'outward Marks of our internal *Ideas*'.

[13] Nuchelmans, *Judgement and Proposition*, p. 77.

assimilates particles, verbs, and copula to syncategorematic terms, markers of operations and not conceptions.[14] If one objects that the functional difference between the marks of affirmation and negations and interpropositional particles is too strong to allow to merge them into one general category, it is sufficient to my argument that there is an analogy between the two insofar as they all mark acts of the mind.[15]

Secondly, and above all, Locke speaks of *is* and *is not* as 'general marks of the Mind, affirming or denying'. The use of the qualifier *general* seems to me to conform to the hypothesis according to which various particular acts are so marked by the terms. In the context, Locke says indeed that there is a diversity of ways of signifying particular actions of the mind relative to particular circumstances, this diversity being in part covered by affirmation and negation, and in part by the diversity of particles.

The proposition is thus made up of terms by which the mind signifies ideas, and by terms, particles, or affirmation or negation, which signify the acts of the mind with regard to its ideas. Besides, particles also serve to mark the various ways in which the mind, 'in declaring its Sentiments to others', establishes connections between propositions. Locke has in mind coordinating conjunctions or connectors like 'therefore', which are crucial to the formation of discourse. Particles do not thus signify conceptions, but actions or sentiments of the mind concerning conceptions. But particles, like the other terms, are components of a proposition or of a series of propositions. The act of the mind is marked by a term that is internal to the proposition, or (in the case of conjunctions like *and*, *therefore*, etc.) interpropositional. Moreover, this term has most often also a structuring role, constituent with regard to this proposition or series of propositions: it links ideas to form a proposition or propositions to form a discourse. It is very clear in the example that Locke gives about the particle *but*, on which he outlines a work that Leibniz, in the parallel section of his *New Essays*, greets as very promising if it was followed in detail, for it shows us the very great richness and fine diversity (in degrees) of 'turns of the mind' ('*tours de l'esprit*'), as Leibniz says in his comments on Locke's remark: 'It seems to me to intimate several relations, the Mind gives to the several Propositions or Parts of them, which it joins by this monosyllable' (E III. vii. 5). It is clear here that the same particle, taken in just one of its significations (for it can have a lot of others), marks simultaneously these two actions: it indicates the relation of the mind with the content that is proposed and it signals the articulation of this content. Now, the fact that Locke tackles the status of the copula, which 'marks'

[14] With regard to 'particles', Locke does not emphasize any difference between '*Prepositions* and *Conjunctions*' (E III. vii. 3): both should be viewed as '*marks of some Action, or Intimation of the Mind*' (E III. vii. 4). By contrast, in the Port-Royal *Grammaire générale et raisonnée* (II. i), the general division of words according as they signify 'the objects of our thoughts' or 'the form or manner of our thoughts' applies specifically to 'particles'. For Arnauld and Lancelot, only conjunctions are in the same class of words as verbs and the copula: prepositions signify conceptions, not the operation of joining or disjoining conceptions. Thanks to Martine Pécharman for pointing that out to me.

[15] I try thus to answer Martine Pécharman's objection (personal communication) that Locke makes a sharp distinction between the functions of particles in the proper sense and that of the copula.

affirmation or negation, in the chapter he devotes to 'particles' (E III. vii), encourages us to explore similarities with Port-Royal.

For this combination of predication and assertion under the same term is also a thesis of *La Logique ou L'Art de penser*.[16] The following passage could have been read by Locke:

It is not enough to conceive these two terms, but the mind must connect or separate them. As we have already said, this action of our mind is indicated [*marquée*] in discourse by the verb 'is', either by itself when we make affirm an affirmation, or with a negative particle in a denial. Thus when I say, 'God is just', God is the subject of this proposition, 'just' is the attribute; and the word 'is' indicates [*marque*] the operation of the mind that affirms, that is, that connects the two ideas '*God*' and '*just*', as belonging together.[17]

Let us note the use of the term 'to separate' ('*séparer*'), which designates the work of the negation in this context. This term will be abundantly employed, later, in II. xix, 'On the nature of negative propositions'. Locke uses the corresponding English term.[18] We also see that, in particular contexts, the connection between ideas, or their separation, is at the same time an assertion of the affirmative proposition or of the negative proposition. As these ideas fit one to the other, when I say that 'God is just', I do not simply predicate *just* of *God*, I agree to the truth of this proposition.[19]

Does it mean that Arnauld and Nicole confuse assertion with predication? No, even if they think that in many cases it is the same syncategorematic term that simultaneously signifies them, so that we can speak indistinctly of 'affirmation'. In the other cases, the two functions diverge, as in the use of affirmation as a simple connection without assertion. This depends on the context of the use and in particular on the intention of the speaker and on the success in the expression of this intention. In the following passage (present in editions 1 to 4, so that Locke, who had the fourth edition

[16] On Locke's view on particles and the copula and its deep affinity with Port-Royal, see Nuchelmans, *Judgement and Proposition*, pp. 145–6.

[17] Arnauld and Nicole, *Logic or the Art of Thinking*, II. iii, p. 82.

[18] I thank Éric Marquer for reminding me that the vocabulary of 'separation', to account for negation, is not at all specific to Locke and Port-Royal, but is traditional and dates back to Aristotle's *De Interpretatione*.

[19] The understanding of 'affirmation' that Walter Ott ascribes to Arnauld is oversimplified: 'To say that I affirm *x* of *y* is simply to say that I am thinking of *x* and *y* in a particular manner: this is what it is to make a judgement.... We are not entitled to infer that all judgements are asserted. For assertion, unlike judgement in the Port-Royalian sense, must have a proposition as its object: one asserts propositions. But one does not judge propositions, in the sense Arnauld *et al.* specify. Their account so far says nothing at all about the assertive force of the resulting proposition' (Ott, *Locke's Philosophy of Language*, pp. 45–6). Ott is right to point out that, for Arnauld, to judge is to produce the proposition, to constitute it, and not to operate on it from the outside. But because in contemporary conceptions the object of assertion must be a proposition, he concludes that the assertion cannot be a function of judgement, merely because of the fact that, for Port-Royal, judgement does not have the proposition as its object. Thus Ott takes for granted that assertion requires a propositional object, a premise that is foreign to Locke's epistemology. The problem with this interpretation is also that it neglects the fact that, for Arnauld, judgement has an assertive force in some contexts. For a persuasive criticism of Ott's understanding of judgement as a subpropositional act in Locke, see Marušić, 'Propositions and Judgments in Locke and Arnauld: A Monstrous and Unholy Union?'

on his shelves, could have had access to it), the authors of the *Logic* highlight the assertive role that the verb acquires in certain contexts and the way in which it is affected by the complexity of propositions:

Besides propositions whose subject or attribute is a complex term, others are complex because they contain terms or subordinate propositions that affect only the form of the proposition, that is, the affirmation or negation expressed by the verb. When I say, for example, 'I maintain that the earth is round', 'I maintain' is only a subordinate proposition that must be part of something in the principal proposition. Yet it is obvious that it is part neither of the subject nor of the attribute. For it changes nothing at all in them, and they would be conceived in exactly the same way if I simply said, 'the earth is round'. And so this affects only the affirmation, which is expressed in two ways, one in the usual way by the verb 'is': 'the earth is round'; the other more explicitly by the verb 'I maintain'.[20]

A clue to our initial problem is thus to be found in the area of grammar and in the sources used by Locke, whether modern or medieval, which are the same as those of the logicians and grammarians of Port-Royal. We have, however, good reason to think that Locke did not have access to the chapter in Arnauld and Nicole's *Logic* dedicated to verbs, since this chapter does not appear before the fifth edition.[21]

Locke does not elaborate on the way in which, in practice, in the understanding of the use of particles, we understand the acts of the mind that are thus signified, but I believe that he had a contextualist theory in mind. We can notice it when Locke is enthusiastic about the immense diversity of signification of one and the same particle in Hebrew, which serves as an argument for supporting the hypothesis of an even greater diversity of attitudes or 'postures' of the mind:

Neither is it enough, for the explaining of these Words, to render them, as is usual in Dictionaries, by Words of another Tongue which came nearest to their signification; for what is meant by them, is commonly as hard to be understood in one, as another Language. They are all *marks of some Action, or Intimation of the Mind*; and therefore to understand them rightly, the several views, postures, stands, turns, limitations, and exceptions, and several other Thoughts of the Mind, for which we have either none, or very deficient Names, are diligently to be studied. Of these, there are a great variety, much exceeding the number of Particles, that most Languages have, to express them by; and therefore it is not to be wondred, that most of these Particles have divers, and sometimes almost opposite significations. In the Hebrew Tongue, there is a Particle consisting but of one single Letter, of which there are reckoned up, as I remember, seventy, I am sure above fifty several significations. (E III. vii. 4)

[20] Arnauld and Nicole, *Logic or the Art of Thinking*, II. 8, pp. 94–5. I agree with Jennifer Smalligan Marušić's reading of 'is' and 'I maintain' in this passage as expressing, for the authors of the *Logic*, affirmation in the 'doxastic sense', not in the 'subpropositional sense'. See her 'Propositions and Judgments in Locke and Arnauld: A Monstrous and Unholy Union?', pp. 268–9.

[21] On verbs as markers of affirmation in the *Logic*, see Pariente, *L'Analyse du langage à Port-Royal*, pp. 126–8, 261–3. On Locke's access to the *Logic*, see van der Schaar, 'Locke and Arnauld on Judgment and Proposition', p. 332.

Locke then takes the example of *but*. With regards to this, Walter Ott, in his chapter on 'Particles and propositions', writes that Locke distinguishes five distinct uses.[22] But he omits to mention what follows in E III. vii. 6, which is crucial: 'To these, I doubt not, might be added a great many other significations of this Particle, if it were my business to examine it in its full latitude, and consider it in all the places it is to be found.'

A point to consider: the same verbal particle can have a great variety of significations that are not conceptions but acts of the mind with regard to these conceptions. In the absence of further textual evidence, my best argument amounts to saying, against Owen, that if, for Locke, it is not improbable that a particle in Hebrew made up of a single letter has more than fifty different significations, it is not unreasonable to suppose that, analogously, affirmation or negation may mark a couple of different acts of the mind.

4. Conclusion

If we accept that for Locke as for Port-Royal a great diversity of acts of the mind is marked by syncategorematic terms, we render plausible the thesis whereby terms such as affirmation and negation can mark various distinct acts of the mind. We have seen that Locke explicitly allows it for particles that, by connecting propositions, constitute a 'discourse'. The price to pay for such an interpretation is that the signification of syncategorematic terms is very dependent upon the context in which they are expressed and in particular upon the intention of the speaker who uses them. This price is perfectly acceptable since this dependence with regard to context and intentions is coherent with the conception of signification employed by both Locke and Port-Royal.

It is therefore not very surprising, although it is regrettable, that a philosophy centred on ideas, which are themselves signified by categorematic terms, and on operations that the mind conducts on these ideas, seems to pay insufficient attention to the importance of what syncategorematic terms mark, since they mark the very operations of the mind in their diversity. But it would be unfair to reproach Locke for not having developed sufficiently this part of the *Essay*, since he explains to his readers that this research should be carried out by each of them, through reflection on their own operations. Indeed, Locke is obviously convinced that what we say about syncategorematic terms is very poor in comparison with the variety of mental postures that they signify, and that this variety is only accessible by introspection or observation of conversations:

> ...But though *Prepositions* and *Conjunctions*, etc. are names well known in Grammar, and the Particles contained under them carefully ranked into their distinct subdivisions; yet he who would shew the right use of Particles, and what significancy and force they have, must take a

[22] Ott, *Locke's Philosophy of Language*, p. 40.

little more pains, enter into his own Thoughts, and observe nicely the several Postures of his Mind in discoursing. (E III. vii. 3)

It is tempting to launch a similar invitation to scholars who would like to make us believe that, according to Locke, when we use affirmation or negation, we mark each time only a single posture of the mind. Once we turn our attention to the philosophy of language it is very difficult to accept.[23]

References

Arnauld, Antoine, and Nicole, Pierre. *Logic or the Art of Thinking*, trans. Jill Vance Buroker. Cambridge: Cambridge University Press, 1996.

Ayers, Michael. *Locke. Volume I: Epistemology*. London: Routledge, 1991.

Jaffro, Laurent. 'Pourquoi l'objet de l'assentiment est-il, selon Locke, la proposition verbale ?' in L. Jaffro, ed., *Croit-on comme on veut? Histoire d'une controverse*. Paris: Vrin: 2013: 147–66.

Locke, John. *Vindications of The Reasonableness of Christianity*, ed. Victor Nuovo. Oxford: Oxford University Press, 2012.

Marušić, Jennifer Smalligan. 'Propositions and Judgments in Locke and Arnauld: A Monstrous and Unholy Union?', *Journal of the History of Philosophy*, 52/2 (2014), 255–80.

Nuchelmans, Gabriel. *Judgement and Proposition: From Descartes to Kant*. Amsterdam: North-Holland Publishing Company, 1983.

Ott, Walter R. *Locke's Philosophy of Language*. Cambridge: Cambridge University Press, 1994.

Owen, David. 'Locke and Hume on Belief, Judgment and Assent', *Topoi*, 22/1 (2003), 15–28.

Pariente, Jean-Claude. *L'Analyse du langage à Port-Royal: six études logico-grammaticales*. Paris: Éditions de Minuit, 1985.

Schaar, Maria S. van der. 'Locke and Arnauld on Judgment and Proposition', *History and Philosophy of Logic*, 29/4 (2008), 327–41.

[23] The first draft of this paper was translated into English by Raphaële Espiet-Kilty.

11

Cartesian Logic and Locke's Critique of Maxims

Andreas Blank

1. Introduction

In the *Essay Concerning Human Understanding*, Locke argues against the view that 'maxims' such as 'The whole is greater than its (proper) parts' or 'It is impossible for something to be and not to be' can serve as the foundation of knowledge. As Michael Ayers summarizes Locke's argument: 'Less general propositions logically derivable from received maxims are, if the maxims are true, just as evident in themselves and "easier and earlier apprehended" in the method of teaching.'[1] Locke's alternative proposal is that maxims have only very limited functions, either for the purposes of giving an orderly exposition of what we already know, or in polemical contexts: If an opponent asks indefinitely for further explanations, we can bring the dispute to a close by invoking truisms that do not need any explanation. In contrast to other parts of the *Essay*, the chapter on maxims has not provoked strong objections among Locke's commentators. Locke's criticism seems to be cogent, and his own view plausible.

Things change, however, if one draws parallels between Locke's critique of maxims and the critique of maxims found in Cartesian logicians such as Johannes Clauberg, Antoine Arnauld, and Pierre Nicole. To be sure, Locke's criticism is directed at the 'Schools'[2] and 'Scholastick Men'.[3] Nevertheless, Ayers points out the close similarities between Locke's argumentative strategy concerning maxims and the objections against the Ramist logic of 'topics' or 'loci' in Arnauld and Nicole's *La Logique ou L'Art de penser* (1662):

> In a passage from Virgil used by Pierre de la Ramée and discussed in the Port Royal *Logic*, a speaker says . . . 'Kill me, not him: I led him to do it'. This was classified by Ramée as an argument from the efficient cause, and for Arnauld and Nicole it illustrates the uselessness of the 'topics' for discovery.[4] It would be absurd to suppose that the speaker might have worked studiously

[1] Ayers, *Locke*, I, p. 92. [2] E IV. vii. 8, p. 595. [3] E IV. vii. 11, p. 598.
[4] Arnauld and Nicole, *La Logique*, III, 17, pp. 232–3.

through a list of topics in order to hit upon such a reason for requesting to be killed. Reasons come naturally to us as the product of common sense and knowledge of the subject: logicians' classification comes later.[5]

In Ayer's view, the authors of the Port-Royal *Logic* emphasize the epistemological priority of our knowledge of particulars and the inability to generate new knowledge through invoking general precepts. As we will presently see, a similar line of argument against Ramism can also be found in the first Cartesian logician, Johannes Clauberg. Arnauld and Nicole discovered Clauberg's *Logica vetus et nova* (1654) only after the printing of the Port-Royal *Logic* started. Nevertheless, Arnauld and Nicole took the opportunity to add some material to the later parts of *La Logique*, using Clauberg's categorization of grammatical, logical, and metaphysical *loci*.[6] Clauberg was involved in intense scholarly and personal struggles with his Ramist colleagues during his short stay at the Academy of Herborn (1649–51)—struggles that eventually led Clauberg to leave Herborn and to accept a position at the newly founded Academy of Duisburg.[7] Like the Port-Royaliens, Clauberg understood Cartesian methodology as an alternative not only to Scholastic but also to Ramist logic. As it turns out, Clauberg raised objections against the usefulness of maxims that are similar to Arnauld and Nicole's and, in part, were directed against the Ramist strands in the thought of one of his Herborn colleagues, Cyriacus Lentulus.

Does it matter whether criticism is directed against the Scholastic conception of maxims or the Ramist conception? It matters a great deal. While the examples of maxims discussed in both traditions are to a large extent similar, and there was a wide agreement about the truth of such general propositions, the differences between the Scholastic and Ramist conception of maxims concern questions regarding the function of maxims in human reasoning. Consider the entry on maxims in Rudolph Goclenius' *Lexicon philosophicum* (1613), a work that provides illuminating insights into early seventeenth-century usage of philosophical terms. Goclenius draws a sharp distinction between his own, late Aristotelian conception of maxims, and the Ramist conception. He defines: 'Maxims are common axioms... they are general rules, comprising the entirety of those things that behave in the same way.'[8] As he explains, in Aristotelian

[5] Ayers, *Locke*, I, pp. 89–90.

[6] Arnauld and Nicole, *La Logique*, III, 18, p. 237: 'En voicy une qui paroit assez commode d'un Philosophe Allemand fort judicieux & fort solide nommé Clauberge, dont la Logique m'est tombée entre les mains, lors qu'on avoit déja commencé à imprimer celle ci.' On the respective responsibilities of Arnauld and Nicole for various portions of *La Logique*, see McKenna, 'La Composition de la Logique de Port-Royal'. Since the passages on maxims are not contained in the so-called '*Manuscrit Vallant*' (BNF, fr. 19 915), which is attributed to Arnauld alone, they might be supposed to be added either by Arnauld and Nicole together, or by Nicole alone. (Heartfelt thanks to Martine Pécharman for pointing this out to me.) For a reproduction of the '*Manuscrit Vallant*', see Dominique Descotes's critical edition of the 1664 version of *La Logique*, pp. 685–801.

[7] On Clauberg's tensions with the Herborn Ramists, see Bohatec, *Die cartesianische Scholastik in der Philosophie und reformierten Dogmatik des 17. Jahrhunderts*, pp. 57–60; Verbeek, 'Johannes Clauberg: A Bio-bibliographical Sketch', pp. 186–7.

[8] Goclenius, *Lexicon philosophicum*, p. 674: '*Maximae sunt communia Axiomata... sunt Generales Regulae, summas earum rerum, quae eodem se habent modo, complectentes.*'

usage axioms are understood in a proof-theoretic context as the propositions that serve as starting points for demonstrations but are not demonstrable themselves.[9] Such a conception of axioms can be found, for example, in Marin Mersenne's *La Verité des sciences* (1625), where propositions such as 'The whole is greater than its parts' and 'Every composite has parts' are characterized as 'maxims and antecedents of our demonstrations' (*maximes, & antecedens de nos démonstrations*).[10] However, Goclenius also mentions a second, alternative conception of maxims. As a spokesman for this second conception, he mentions the German Ramist Friedrich Beurhaus (1536–1609). Beurhaus defines: 'Maxims are documents of an art, either the precepts themselves or at least their outcome and consequence.'[11] Moreover, he calls a maxim in logic a 'locus communis'.[12] These quotations indicate that, according to the Ramist view, maxims are not part of proofs but rather part of the logic of 'places'. Moreover, maxims stand in a complex relationship to an 'art'—a discipline such as logic, rhetoric, or jurisprudence—a relationship that combines two aspects: maxims have a prescriptive function and, in this sense, are 'precepts' that tell us how to reason correctly; but our practices of reasoning do not depend on the explicit formulation of maxims and, in this sense, maxims 'document' rather than constitute an art. The distinction between a proof-theoretic and a topical conception of maxims will be crucial for understanding both the merits and the shortcomings of Locke's critique of maxims.[13]

Widening the context of Locke's critique of maxims, as I propose to do here, clarifies what is going on in Locke's chapter on maxims: Clauberg and the authors of the Port Royal *Logic* develop a line of argument against the Ramist conception of maxims, and Locke applies a similar line of argument against a different target: the Scholastic conception of maxims (or what Locke took this conception to be). This shift of the object of criticism obviously may bring epistemic benefit: Even if Clauberg, Arnauld, and Nicole may have fallen short of refuting the Ramist conception of maxims, the same line of argument may prove successful in refuting the Scholastic conception of maxims. Nevertheless, widening the context of Locke's critique of maxims also indicates what is going wrong in Locke's chapter on maxims. The Cartesian criticisms miss the point of the Ramist conception of maxims. According to the Ramists, maxims do not contribute to

[9] Ibid., p. 136.

[10] Mersenne, *La Verité des sciences*, p. 192. On the Scholastic background of Mersenne's thought, see Dear, *Mersenne and the Learning of the Schools*.

[11] Goclenius, *Lexicon philosophicum*, p. 674: '*Maximae sunt artis documenta, praeceptaque ipsa vel saltem eorum fructus & consectaria*.'

[12] Goclenius refers the reader to Beurhaus, *Paedagogiae logicae... pars secunda, editio locupletior*, p. 324.

[13] Logicians to be found in Locke's library include Clauberg and the authors of the Port-Royal *Logic* (although none of them is included in the list of annotated texts). But even more prominent than works from the Cartesian tradition were writings of the protagonists of the new Baconian science: besides works by Boyle, Sydenham, and Hooke, there are numerous volumes by William Molyneux, Thomas Willis, Thomas Burnet, Joseph Glanvill, Claude Perrault, and others (see Harrison and Laslett, *The Library of John Locke*). Locke does not seem to have owned any works by Petrus Ramus and his literary associate, Omer Talon, or any works by the prominent Ramist critic of Baconian methodology, Sir William Temple. On the absence of the most prominent anti-Baconian works in Locke's library, see Wood, *The Politics of Locke's Philosophy*, pp. 65–7.

increasing our knowledge about the world. Rather, according to the Ramist conception, maxims are descriptive in the sense that they contribute to our knowledge about our own argumentative practices: They make something explicitly known that was known only in an implicit way before. To be sure, it was not the aim of Locke to refute the Ramist conception of maxims. But he did not take it into consideration, either. Due to its strong consilience with the arguments of Clauberg, Arnauld, and Nicole, Locke's criticism of the Scholastic conception of maxims leaves the Ramist alternative untouched. Hence, because Locke does not take the Ramist alternative into account, his conclusion about the limited functions of maxims remains inconclusive. This point is interesting because it indicates a sense in which philosophy could be descriptive—a sense that the Ramists have captured and Locke has missed.

2. What is Going On?

Let us begin with Clauberg's critique of maxims. Clauberg was an eclectic who tried to combine elements of Aristotelian philosophy with elements from Cartesian methodology. With respect to ontology, Clauberg was a Scholastic who thought that the mind of God guarantees the possibility of the correspondence between signs and what is signified.[14] With respect to logic, however, he tried to strike a balance between Aristotelian logic and the logical implications of the Cartesian method. He accepted Descartes's epistemology of clear and distinct ideas and the view that correct reasoning depends on being attentive with respect to one's own ideas.[15] He defended the Cartesian method of doubt—a topic to which he devoted a whole book.[16] Also, he shared Descartes's view that some errors arise through the persistence of errors from childhood.[17] Clauberg adds to Descartes's view of the origin of errors Bacon's conception of 'idols', especially the illusions which arise out of the misleading usage of words and the illusions which arise out of interactions with other human beings.[18] Most importantly, as far as the logic of invention goes, Clauberg held that Descartes's methodological precepts are all that the human mind needs. As Paul Schuurman puts it, Clauberg 'defines logic as an art that should not concern itself with arid sophistry but that should rather help us direct our mind in the acquisition of new knowledge... Clauberg takes care to stress

[14] On Clauberg's ontology, see Viola, 'Scolastica e cartesianesimo nel pensiero di J. Clauberg', pp. 256–64; Trevisani, 'Johannes Clauberg e l'Aristotele riformato'; Savini, *Johannes Clauberg:* Methodus Cartesiana *et ontologie.*

[15] Clauberg, *Logica vetus et nova*, I, ii, §16; I, iii, §1; see Savini, 'L'Insertion du cartésianisme en logique', pp. 76–80.

[16] Clauberg, *Initiatio philosophi sive dubitatio cartesiana.*

[17] Clauberg, *Defensio cartesiana*, chs. xxxii–xxxiv; Clauberg, *Logica vetus et nova*, Prolegomena, ch. 2, §§20 and 24; Descartes, *Meditationes de prima philosophia* I, AT 7, p. 17; Descartes, *Sextae responsiones*, AT 7, p. 441; Descartes, *Principia*, I, art. 1, AT 8A, p. 5; I, art. 72, AT, 8A, p. 36; Descartes, *Discours*, II, AT 6, p. 13; see Savini, 'L'Insertion du cartésianisme en logique', pp. 80–5.

[18] Clauberg, *Logica vetus et nova*, Prolegomena, ch. 4; Ibid., IV, ii–iv; see Savini, 'L'Insertion du cartésianisme en logique', pp. 85–6.

that Descartes did not dismiss the use of Aristotelian syllogisms out of hand, but rather inveighed against a dialectic that limited itself to disputations and sterile *loci topici*.'[19]

In his rejection of the use of *loci* such as maxims, Clauberg goes beyond Bacon's critique of 'idols'. Bacon had thought that maxims are fine as far as the rules of discourse go. In fact, in one of his early works he had written extensively about the role of maxims in juridical argumentation, and he came back to the role of maxims in negotiations in *De dignitate et augmentis scientiarum*.[20] Moreover, the fact that they do not express any insight that we did not possess before, in his view, does not impair the value of such maxims. As he argues in *The Advancement of Learning*, 'invention of sciences' differs profoundly from 'invention of speech':

The *Invention* of speech or argument is not properly an *Invention*: for to *Invent* is to discover that, we know not, & not to recover or resummon that which wee already knowe; and the use of this *Invention*, is no other; *But out of the Knowledge, whereof our minde is already posseest, to draw foorth, or call before us that which may bee pertinent to the purpose, which wee take into our consideration*. So as to speake truly, it is no *Invention*; but a *Rememberance* or *Suggestion*, with a Application...[21]

In fact, even if Bacon held 'anticipations of the mind' to be useless for the purposes of science,[22] he did not hesitate to apply one juridical maxim to the mode of exposition chosen in the *Novum organum* when he surmises that there is no possibility of argument with alternative theories of scientific method 'since we do not agree about the principles nor about the proofs'.[23] Consequently, Bacon had urged the peaceful coexistence of two distinct types of invention, one relevant for science, and one relevant for linguistic expression.[24]

By contrast, Descartes argues against the usefulness of maxims even for the art of disputation. In the *Discours de la méthode*, Descartes remarks that 'many people use the art of Lull only to quarrel abundantly and without any judgement about things that they do not know'.[25] Likewise, in his *Epistola ad G. Voetium* (1643), he rebukes 'childish dialectics' (*puerilis dialectica*):

And even those who have a quick and hot imagination but *no judgement*, just like children, can acquire within a few days the usage of this great art: for most easy are *similarities, differences, contraries, circumstances, antecedents, consequents*, and others of this kind, which are commonly counted among the *loci*...All objections and replies that, since they are unable to find them out by means of the internal consideration of the difficulty itself, they derive from the consideration

[19] Schuurman, *Ideas, Mental Faculties, and Method*, p. 63. On Clauberg's replacement of logic of loci by Descartes's methodological precepts, see Trevisani, *Descartes in Germania*, pp. 73–8. Clauberg's interest in Cartesian methodology, of course, is compatible with his view that the Cartesian methodology has to be supplemented by other rules for other purposes. On Clauberg's theory of interpretation, see Lagrée, 'Sens et verité chez Clauberg et Spinoza'.

[20] Francis Bacon, *Maxims of the Law*; *De dignitate et augmentis scientiarum* VIII, 1–2, in *Works*, vol. 1, pp. 745–91.

[21] Bacon, *The Advancement of Learning*, p. 111.

[22] Bacon, *Novum organum*, I, xxxiii.

[23] Ibid., I, lxi.

[24] Ibid., preface.

[25] AT 6, p. 17.

of these external loci, are almost always inept and futile...And for this reason it is *most detrimental*, not only for adults but even more so for adolescents, who when they get used to it and become arrogant through the opinion of the doctrine acquired with it, they plainly corrupt their natural reason, which otherwise would have become mature with age.[26]

In his *Defensio Cartesiana* (1652), Clauberg responded to some of the controversies provoked by Descartes's writings on methodology. One early work that came out with criticisms of Descartes's dismissal of the art of invention was the *Methodi cartesianae consideratio theologica* (1648) by the Dutch Reformed theologian Jacobus Revius. Most importantly for the present concerns, Revius objects that, in the cited passage from the letter to Voetius, Descartes did not prove that dialectics corrupts natural reason and suggests that art does not corrupt, but rather perfects nature.[27] Cyriacus Lentulus, professor of politics at the Academy of Herborn, took up a similar line of argument in his *Nova Renati Descartes sapientia* (1651). Lentulus maintains that, in the cited passage from the *Discours*, Descartes suppresses topics and the art of invention

[w]hich in conjunction with the forms of argumentation helps in investigating things that we do not know, for the reason that it is no little help in giving to others an exposition of what we know...Of course, logic alone does not promise knowledge of things: just as an instrument does not work unless some material, say wood or cement, is provided, and the instrument is equipped with the mind and also the hand of the architect. Nevertheless, in order to reach an affirmation concerning things that we are doubtful about, from two more known propositions a third is deduced by means of necessary consequence: and the premise itself requires the art of topics.[28]

In his response to these objections, Clauberg explains that Descartes does not disapprove of the investigation of causes and effects, but has a low opinion of arguments that are derived from general maxims and not from the internal nature of things.[29] Clauberg

[26] AT 8B, pp. 50–52: '*Et quidem ii qui promtae et calidae sint imaginationis, sed nullius iudicii, quales pueri esse solent, possunt intra paucos dies magnum istius artis usum acquirere; perfacile enim similitudines, differentias, contraria, adiuncta, antecedentia, consequentia, et reliqua ejusmodi, quae vulgo in topicis recensentur...Omnes...objectiones et rationum solutiones, quae cum ex intima ipsius difficultatis consideratione erui non possunt, ab externa locorum istorum inspectione petuntur, fere semper sunt futiles et ineptae...Et ideo damnosissima est, non tantum adultis, sed praecipue junioribus, qui dum ei assuescunt, et ex opinione doctrinae per ipsam acquisita, superbiunt, rationem suam naturalem, quae alioquin aetate posset maturescere, plane corrumpunt.*' On the Utrecht controversy, see Verbeek, *Descartes and the Dutch*, pp. 13–33; van Ruler, *The Crisis of Causality*.

[27] Revius, *A Theological Examination of Cartesian Philosophy*, p. 130: '*Qui hoc dicit, nec probat, pro maledico calumniatore traduci meretur. Quis enim non videt naturam arte expoliri, non autem corrumpi?*' See Savini, *Johannes Clauberg:* Methodus Cartesiana *et ontologie*, pp. 105–9.

[28] Lentulus, *Nova Renati Descartes sapientia*, pp. 50–1: '*[Q]uae una cum forma argumentationis ad ea quae ignoramus investiganda prodest, eo ipso argumento, quod non parum prosit in iis quae novimus aliis exponendis...Rerum quidem cognitionem Logica sola non promittit: ut nec instrumentum operatur nisi materia, puta ligno aut caemento, praebita, & architecti mente vel etiam manu ornatum. Ut tamen ad eorum, de quibus dubitatio instituitur, asseverationem perveniamus, e duobus notioribus tertium de quo quaerebatur necessaria illatione concluditur: & praemissa ipsa arte topica indagat.*' See Savini, *Johannes Clauberg:* Methodus Cartesiana *et ontologie*, pp. 113–16.

[29] Clauberg, *Defensio cartesiana*, p. 120: '*Non improbat caussarum, effectorum &c. investigationem...sed argumenta ab externis illis locis causae, effecti &c. vi maximarum istarum nimis generalium & non ex intima rei natura petita tanquam Philosophus parvi pendet.*'

also adds an argument for why dialectics is supposed to corrupt our natural reason: '[Descartes] proves this from the consideration that for this reason they are distracted from the internal consideration of things, and they are led to say many things without judgement about things of which they do not know the natures and properties.'[30] In the *Logica vetus et nova*, Clauberg defends the Cartesian view with respect to maxims and holds that logical and metaphysical maxims are useless and misleading even as rules of disputation.[31] He offers the following assessment:

Even if several maxims, namely those that are derived from definition, from cause, from the whole, etc., are such that the mind cannot doubt their truth when expressed generally, their *application* to a given question remains controversial, namely, whether the definition that is proposed by the dialectician is the true definition of the thing, whether this is the cause, and this the effect, whether this is a whole or a part; unless the adversary concedes these, or the participant in the disputation is able to deduce these from the consideration of the thing itself and its internal nature, his rule is of no use at all; but if he is able to deduce them from there [the thing itself and its internal nature], he will not produce clouds by arguing from common *loci*, but he will demonstrate [something] from the nature of the thing that is known to him.[32]

According to Clauberg, the problem with maxims is not that they are not true. Rather, even in the case of maxims that are certain we face a problem of how to apply them to a given case. Clauberg's argument is that, if we manage to apply them to a given case in a way that yields a correct result, this is due to our having analysed this particular case correctly. For example, only when we have correctly identified a cause of a given effect and a cause of this cause, can we apply one of the causal maxims that Clauberg mentions before—'The cause of a cause is a cause of the effect'—to the given case. But then, Clauberg concludes, the application of maxims is secondary to the analysis of particular cases, and our knowledge of particulars in no way depends on maxims. For this reason, Clauberg assigns only a very limited value to the apparatus of dialectics. He concedes that the tools of dialectics may be useful for illustrating particular cases and in popular conversation; however, he holds that dialectics is of no use when it comes to providing

[30] Ibid., p. 121: '*Hoc inde probat, quia ista ratione ab intima rerum consideratione abducuntur, ac de iis quarum naturas & proprietates nesciunt multa sine judicio loqui adsuescunt.*'

[31] See Bohatec, *Die cartesianische Scholastik*, pp. 89–90. As examples of logical loci, Clauberg mentions the following: 'What is universally affirmed or negated about the genus, is also affirmed or negated about the species;' 'If the genus is denied, also the species is denied;' 'If the definition is affirmed or negated with respect to something, also what is defined is affirmed or negated with respect to something;' 'What is affirmed or negated with respect to the definition, is also affirmed or negated with respect to the defined.' As examples of metaphysical loci: 'The nature of the cause determines the nature of the effect;' 'What is the cause of a cause is also a cause of what is caused;' 'If it is assumed that the effect is given, it is assumed that an efficient cause exists or has existed' (Clauberg, *Logica vetus et nova*, p. 193).

[32] Clauberg, *Logica vetus et nova*, pp. 194–5: '*Etiamsi nonnulli canones, videlicet qui a definitione, a causa, a toto, &c. petiti, tales sint, ut de eorum generatim enunciatorum veritate mens dubitare nequeat, de eorum tamen applicatione ad quaestionem res manet controversa, nempe an ea, quam ponit Dialecticus, sit vera illius rei definitio, an ea causa, an id effectum, totum, pars, quod nisi vel ultro concedat adversarius, vel ex rei ipsius consideratione & natura intima educere possit Disputator, ne quidem hilum proficiet sua regula; sin possit inde deducere, non nugabitur sic disputando e loco communi, sed demonstrabit e natura rei sibi perspecta.*'

evidence concerning the facts, or teaching academic disciplines, or engaging 'in private and solitary demonstrations'.[33] As Clauberg sums up, maxims pertain to the popular mode of discourse but not to the scientific mode of discourse (*exotericus* in contrast to *acroamaticus*).[34]

Similar reservations about the usefulness of maxims can be found in the work of the Port-Royaliens. As Bernard Roy has put it: 'Arnauld and Nicole distrusted the mechanistic aspect of the topics and places'—the view that considering maxims and other *loci* would be sufficient or even necessary for finding new arguments.[35] In the chapter on maxims, Arnauld and Nicole are well aware of the reasons that lead Ramus to challenge the Scholastics. While the Scholastics treat the *loci* after having given the rules of argument, Ramus claims against them that one has to explain the *loci* before treating the rules of argument. Ramus' argument is that one first has to find the matter of thought before one can even dream of ordering it.[36] Arnauld and Nicole think that this argument is weak

> because even if it is necessary to find the matter for giving an order to it, it is nevertheless not necessary to learn how to find the material before learning how to order it. For to learn how to order material, it is sufficient to have some general materials that serve as an example; and the mind and the *sensus communis* provide enough such material without any need to apply any art or method. Hence it is true that one has to have some material to apply the rules of arguments; but it is false that it is necessary to find this material by means of the method of *loci*.[37]

According to this line of argument, the human mind has a natural capacity to find new thoughts. Therefore, finding new thoughts does not depend on reflecting on any precepts of dialectics. Arnauld and Nicole even deny that, apart from dialectical exercises, anyone ever applies dialectical rules such as maxims, asking rhetorically: 'Is there a single person who could say truthfully that, when he was obliged to treat some issue, he has reflected on these *loci* and searched there the reasons that were necessary for him?'[38]

[33] Ibid., pp. 195–6: '*Nullumne igitur usum apparatus dialectici esse concedis? Resp. Usum in levioribus praeludiis do, in decretoriis praeliis nego, do ad illustrandum, nego ad evidenter convincendum, do in populari conversatione, nego in disciplinarum traditione, do usum rationibus dialecticis demonstrationi sociatis, nego separatis ac solitariis.*'

[34] Ibid., pp. 136–6. [35] Roy, 'Reasoned Grammar, Logic, and Rhetoric at Port-Royal', p. 134.

[36] Arnauld and Nicole, *La Logique*, III, 17, pp. 232–3: 'Ramus fait une querelle sur ce sujet à Aristote & aux philosophes de l'école, parce qu'ils traitent des Lieux après avoir donné les regles des arguments, & il pretend contre eux, qu'il faut expliquer les Lieux & ce qui regarde l'invention avant que de traiter de ces regles.

La raison de Ramus est, que l'on doit avoir trouver la matiere avant que de songer à la disposer.'

[37] Ibid., p. 233: 'Mais cette raison est très-foible, parcequ'encore qu'il soit necessaire que la matiere soit trouvée pour la disposer, il n'est pas necessaire neanmoins d'apprendre à trouver la matiere avant que d'avoir appris à la disposer. Car, pour apprendre à disposer la matiere, il suffit d'avoir certaines matieres generales pour servir d'exemples; or l'esprit & le sens commun en fournit toujours assez sans qu'il soit besoin d'en emprunter d'aucun art ny d'aucune methode. Il est donc vrai qu'il faut avoir une matiere pour y appliquer les regles des argumens; mais il est faux qu'il soit necessaire de trouver cette matiere par la methode des Lieux.'

[38] Ibid., p. 234: '[Y] en a-t-il un seul qui puisse dire veritablement, que lorsqu'il a été obligé de traiter quelque sujet, il ait fait reflexion sur ces lieux, & y ait cherché les raisons qui lui etoient necessaires ? Qu'on

But Arnauld and Nicole not only think that in fact no one ever applies dialectical rules when forming arguments in real-life situations. Like Clauberg, they also hold that the general principles called 'maxims' are unsuitable for this purpose since they do not lead the human mind to acquire new knowledge. For example, they question that there could be any circumstance in which the principle 'It is impossible for the same thing to be and not to be' could serve to acquire any knowledge.[39] In his view, this distinguishes maxims from the rules of Cartesian methodology, which serve to acquire new knowledge about the mind and the external world.

A similar line of argument is found in Locke's various remarks on maxims, from the early Drafts A and B of the *Essay* (1671) to the *Essay* itself and Locke's additions to later editions of the *Essay*. I do not want to suggest that Locke derived any of his remarks from the work of Clauberg or the Port-Royal *Logic*. Rather, I would like to point out a similarity in argumentative structure. Already in Draft A, it becomes clear that Locke thinks of maxims in a proof-theoretic context when he comments on the principle 'What is is' or 'The same thing cannot be and not be': 'But how it influences the demonstration of the truth of things & upon what ground which perhaps hath by noe body been touchd at shall be shewne . . . amounting to noe more then this that the same word may with great certainty be affirmd of its self without any doubt of the truth of any such propositions & let me adde without any knowledg of any thing.'[40] In the *Essay*, Locke develops a similar line of argument:

> These two general Maxims amounting to no more in short but this, that *the same is the same*, and *the same is not different*, are truths known in more particular instances, as well as in these general Maxims, and known also in particular instances, before these general Maxims are ever thought on . . . There is nothing more visible, than that the Mind, without the help of any Proof, or Reflection on either of these general Propositions perceives so clearly, and knows so certainly, that the *Idea* of *White*, is the *Idea* of White, and not the *Idea* of Blue . . . therefore the truth of no general Proposition can be known with a greater certainty, nor add any thing to this.[41]

Locke applies his view that the evidence of propositions such as 'white is not blue' depends on the perception of our ideas of white and blue not only to empirical ideas but also to mathematical ideas. Already in Draft B he claims that the mind has intuitive knowledge about equality and inequality of numbers by means of the clear and distinct notions it has of them. The certainty of such knowledge does not require having

consulte tant d'Avocats & de Prédicateurs qui sont au monde, tant de gens qui parlent & qui écrivent, & qui ont toûjours de la matiere de reste, & je ne sai si on en pourra trouver quelqu'un qui ait jamais songé à faire un argument *à causa, ab effectu, ab adjunctis*, pour prouver ce qu'il desiroit persuader.'

[39] Ibid., IV, 7, p. 321: '[J]e ne voi point de rencontre où il puisse jamais servir à nous donner aucune conoissance.'

[40] Locke, *Drafts for the Essay Concerning Human Understanding*, Draft A, #28, p. 54.

[41] E IV. vii. 4, p. 593.

recourse to any maxims which, interestingly, Locke in this context also calls 'axioms'.[42] A similar view is expressed in the *Essay*:

As to the *Relations* of Modes, Mathematicians have framed many Axioms concerning that one Relation of Equality. As *Equals taken from Equals, the remainder will be Equals*; which, with the rest of that kind, however they are received for Maxims by the Mathematicians, and are unquestionable Truths; yet, I think, that any one who considers them, will not find, that they have a clearer self-evidence than these, that *one and one, are equal to two*; that *if you take from the five Fingers of one Hand two, and from the five Fingers of the other Hand two, the remaining numbers will be equal.*[43]

Again, Locke thinks about maxims in the proof-theoretic framework of axioms and argues that, due to the epistemological role of clear and distinct ideas, maxims (understood as axioms) cannot convey a greater degree of certainty to propositions about particulars than the self-evident insight into clear and distinct ideas.

In a small note with the heading 'Maximes', written between 1697 and 1699 as an addition to the fourth edition of *Essay* IV. xii. 3, Locke develops this theme further. There, he doubts that a child knows how to use the terms 'part' and 'whole' before she knows how to use the terms 'finger' and 'body'. The implication seems to be that a child is able to apply a maxim such as 'the whole is greater than its parts' only after she has mastered the use of the terms 'whole' and 'part'. But if a child has mastered the terms 'finger' and 'body' earlier than these terms, the insight that the body is greater than a finger cannot be derived from any maxim. Locke adds:

I ask farther when he has got these Names, how is he more certain that his Body is a *whole*, and his Little-Finger a *part*, than he was or might be certain before, he learnt those Terms, that his Body was bigger than his Little-Finger? Any one may as reasonably doubt or deny that his Little-Finger is a part of his Body, as that it is less than his Body.[44]

Hence, our capacity of applying a maxim to a given situation depends on our capacity to apply the terms occurring in the maxim to this situation; but, as Locke holds, we cannot be more certain about the applicability of terms than we are about the relations between our clear and distinct ideas. For this reason, maxims cannot increase the certainty of our particular propositions, and Locke concludes: 'that the Maxim, *The whole is bigger than a part*, can never be made use of to prove the Little-Finger less than the Body.'

Rejecting a proof-theoretic conception of maxims leads Locke to the conclusion 'that these magnified *Maxims*, are not the Principles and *Foundations* of all our other *Knowledge*. For if there be a great many other Truths, which have as much self-evidence as they, and a great many that we know before them, it is impossible they should be the

[42] Locke, *Drafts for the Essay Concerning Human Understanding*, Draft B, #51, p. 158.

[43] E IV. vii. 6, p. 594.

[44] E IV. xii. 3, p. 641; cf. Locke, 'Maxims'. MS Locke, e. 1, p. 182. On the latter text, see Schuurman, 'General Introduction' to his Ph.D. edition of Locke's *Of the Conduct of the Understanding*, pp. 100, 111.

Principles, from which we deduce all other Truths.'[45] In particular, he points out that all the discoveries of Newtonian physics have been made without invoking maxims a single time.[46] On the contrary, he argues that, when maxims are applied to erroneous definitions in natural philosophy, they lead to conclusions that are positively wrong.[47] Consequently, Locke suggests that maxims can fulfil only two very limited functions: they are useful 'in the ordinary Methods of teaching Sciences as far as they are advanced'; and they are useful 'in Disputes, for the silencing of obstinate Wranglers'.[48] Of course, the view that maxims can have expository and polemical functions derives from observations concerning textbook didactics and the practices of academic disputation. However, the stronger claim that these are the *only* identifiable functions of maxims seems to be a direct consequence of his insight that arguing from maxims cannot increase knowledge.

By now, it should be clear what is going on in Locke's chapter on maxims: Locke uses an argumentative strategy that closely resembles the argumentative strategy found in Cartesian critiques of maxims. Even if there is no reason to assume that Locke derived his chapter from the works of the Cartesians, it instantiates an argumentative pattern characteristic of the emerging Cartesian logic: maxims are incapable of increasing our knowledge; hence, any conception that regards maxims as the foundation of science has to be dismissed. This argumentative pattern certainly is attractive. But recall that Goclenius in his *Lexicon philosophicum* distinguished between the Scholastic, proof-theoretic conception of maxims as axioms and the Ramist, topical conception of maxims as general rules of dialectics. Is the Cartesian line of argument cogent when applied against the conception of maxims as *loci*? This question turns out to be surprisingly relevant for assessing the cogency of Locke's argument. This is so because moving from the insight that maxims cannot ground and expand knowledge to the view that the *only* functions of maxims are restricted to expository and polemical contexts depends on the absence of further viable alternatives.

3. What is Going Wrong?

As I said in Section 1, placing Locke's chapter on maxims within the context of Cartesian logic not only makes clear what is going on in this chapter. It also makes clear what is going wrong there. Similarity of argumentative structure, not textual influence, is what matters here. Since Locke's arguments are closely similar to those found in Clauberg, Arnauld, and Nicole, one specific shortcoming of the Cartesian arguments carries over to Locke's. The shortcoming I have in mind is the following: Granted, Ramism is not equally present as a topic in all three philosophers. Arnauld and Nicole attack Ramus

[45] E IV. vii. 10, p. 596.

[46] E IV. vii. 11, p. 599. On Locke's view of the uselessness of maxims for natural philosophy, see Yolton, *The Two Intellectual Worlds of John Locke*, pp. 54–8.

[47] E IV. vii. 12, p. 604.

[48] E IV. vii. 11, p. 600.

head on; Clauberg is involved in an academic controversy with Cyriacus Lentulus, a philosopher influenced by some aspects of Ramism; and Locke simply ignores the Ramists. Nevertheless, as I will argue in this section, the arguments of the Cartesians are insufficient to refute the Ramist view of maxims. Hence, even if Locke's arguments may work well when directed against the Scholastic view of maxims, they are as ineffective to exclude the Ramist alternative as the Cartesian arguments are.

What did Ramus think about maxims, and how does his view of maxims relate to his conception of dialectical invention? According to Ramus, 'Invention is the doctrine of thinking & finding an argument, i.e., a reason suitable to explicate a question.'[49] Invention takes place by using concepts that make such an explication possible. In this sense, he regards categorial concepts as *loci* fulfilling this function in the constitution of discourse. He differentiates them by using the traditional distinction between innate (*insita*) and conventional (*assumpta*) concepts.[50] Among the innate, non-conventional concepts, he counts causal concepts such as those of final, formal, efficient, conserving, and spontaneous cause.[51] Ramus claims, 'We can derive from the common use [*communis usus*] of simple prudence all the testimonies of all the merits of dialectics.'[52] Hence, there is a sense in which dialectical invention does not invent anything that we did not have before. Ramus points out that, when he uses examples from literary and rhetorical works, he understands poets and orators as 'famous and illustrious witnesses of this common sense, and of human prudence'.[53] He also puts it thus: 'The art of dialectics has as its subject and aim to explicate the natural use of reason as it is impressed on great minds, in a certain custom, in perfect examples.'[54]

In Ramus' view, logical and metaphysical maxims are part of dialectics. He criticizes the Scholastic view of maxims as a part of scientific knowledge. In his view, maxims are unsuitable to fulfil the function of providing a foundation of knowledge since they do not guarantee that the assumptions from which we start our argument are not faulty. He recommends: 'When we explore given examples . . . not only the positive sides should be discerned, so that relevant, perspicuous, certain, and coherent things are accepted. But also the vices should be noted, so that irrelevant things are rejected,

[49] Ramus, *Institutionum dialecticarum libri tres*, p. 16: '*Inventio est doctrina cogitandi & inveniendi argumenti, id est, probationis ad explicandam quaestionem idoneae*.' On Ramus' conception of loci, see Robinet, 'La Posterité cartésienne du combat raméen et ramiste'.

[50] Ramus, *Institutionum dialecticarum libri tres*, p. 17.

[51] Ibid., pp. 17–37.

[52] Ibid., pp. 372–3: '*Verum e simplici prudentiae communis usu, qui caeteris in rebus arte nulla comprehensis versatur, omnium dialecticarum laudum testimonia sumpsimus: atque ea quidem quam maxime potuimus insignia. Poetas enim & oratores, id est claros & illustres communis illius sensus, & humanae prudentiae testes in singulis doctrinae locis adhibuimus: qui cum dialecticam non solum in privatis doctorum hominum scholis, verum etiam in iudiciis, in senatu, in concionibus, in theatris, in omnibus denique humanae vitae partibus vigentem florentemque demonstraverint, tum facile singularem hunc rationis fructum, omnium hominum communem esse vindicabunt: popularique illa sua non solum sapientia, verum etiam suavitate & elegantia maiores certe virtutis huius amores excitabunt*.'

[53] Ibid., p. 373.

[54] Ibid., p. 58: '*Dialectica ars subiectum & propositum habet ad explicandum naturalem rationis usum in magnis ingeniis, in certa consuetudine, in perfectis exemplis impressum*.' On the role of natural reason in Ramus' dialectics, see Blank, 'Ramus and Leibniz on Analysis'.

obscure ones explained, ambiguous ones distinguished, false ones negated, and incoherent ones suspended.'[55] As he explains, among false assumptions there are those that take something that is not a cause for a cause, or something that is not an effect for an effect;[56] also, there are errors that are due to the misleading imposition of names[57] or that arise from wrong definitions.[58] In his view, maxims are not the adequate tool to avoid errors concerning particular causes of particular effects, or to avoid a misleading usage of names or erroneous definitions of terms. To avoid such epistemic vices, we have to apply epistemic capacities other than those that can be expressed by means of maxims. Maxims, in Ramus' view, have another function:

> We briefly touched upon those vices which seem to be opposed to the virtues of invention, so that we do not expect anything from maxims, as they are commonly called by the Scholastics. Maxims do not carry with them any new precepts: for they have something of the precepts of an art, such that when a definition or distinction, which is taken for granted for the purposes of argument in order to find a proof for a given question, all precepts of this art, when I give my assent to them, are to be related to the confirmation of human prudence...[59]

Presumably, what Ramus has in mind is the following. Suppose we have identified a particular cause of a given effect, and a particular cause of this cause. Then maxims concerning causes and effects, for example, 'The cause of a cause is a cause of the effect', give us rules about how to relate remote segments of a chain of causes to the effect. Such a maxim does not help us in correctly identifying causes and effects. But it tells us what kinds of arguments concerning causes and effects are valid, once these have been identified. Moreover, maxims of this kind do not tell us anything new. Rather, they make rules explicit that we implicitly always apply, for example when reasoning about causes and effects.

Ramus holds that many examples show that such rules are in fact used by classical authors such as Cicero. He asks us to consider a passage from the *Tusculanae disputationes*, where Cicero suggests: 'when it is said that in vices there is a strong force leading towards a miserable life, isn't it to be said that in virtue there is a strong force leading towards a happy life?' According to Ramus, the maxim that Cicero is applying here is the rule according to which 'contraries are the consequences of contraries.'[60] Of course, Ramus does not claim that Cicero is actually reflecting in an explicit way on this maxim.

[55] Ramus, *Institutionum dialecticarum libri tres*, pp. 306–7: '*In explorandis propositis exemplis...non solum laudes cernentur, ut propria, perspicua, certa, vera, consentanea concedantur. Sed vitia notabuntur, ut aliena reijciantur, obscura explicentur, ambigua distinguantur, falsa negentur, inconsequentia suspendantur.*'

[56] Ibid., pp. 309–10. [57] Ibid., pp. 313–14. [58] Ibid., pp. 315–16.

[59] Ibid., p. 329: '*Breviter attigimus exemplis ea vitia quae virtutibus inventionis opposita esse videbantur, nisi forte de maximis propositionibus, quae vulgo nominantur in scholis, aliquid expectamus: quae tamen nihil habent novae praeceptionis: aliquid enim ex artis praeceptis habent, ut definitionem, aut divisionem, quod pro argumento ad quaestionem probandam sumitur, ut affirmanti mihi artis huius omnia praecepta ad humanae prudentiae confirmationem referenda esse...*'

[60] Ibid., p. 330: '... *"cum faterentur satis magnam vim esse in vitiis ad miseram vitam, nonne fatendum est eandem in virtute esse ad beatam?" Contraria enim contrariorum sunt consequentia.*' See Cicero, *Tusculanae disputationes*, 5. 17. 50.

Rather, his suggestion seems to be that the maxim makes a rule explicit that in an implicit, pre-reflective way structures propositions like the one found in the *Tusculanae disputationes*. In this sense, maxims do not enable the mind to do anything that it was unable to do before. Rather, they explicate the rules that the mind had applied in reasoning before reflecting on its own operations.

The Ramist conception of maxims was by no means absent from early seventeenth-century philosophy in England. Take the case of Sir William Temple (1555–1627), one of the most influential Cambridge Ramists and some-time Provost of Trinity College, Dublin.[61] In his commentary on the two-book Latin version of Ramus' *Institutiones dialecticae*, Temple asks:

> From the observation of which thing did... the precepts of the art of logic flow? Certainly from the observation of natural reason, viz. comprised and expressed in the sayings and writing of other men who are famous in the Republic of Letters, in the opinions and commentaries of poets, orators, philosophers, theologians. For they all have made use of the natural and divine endowed reason, i.e., they made the causes and effects of things evident...[62]

Like Ramus, Temple maintains that there is nothing novel or artificial in the rules of logic. Rather, logic describes what is going on in our natural capacities of reasoning. This is why the study of the most outstanding examples of natural reason, as documented in the works of great poets, orators, etc., is essential: their patterns of reasoning provide the standard against which other arguments can be assessed. The origin of logic in natural reason, in Temple's view, also speaks against the widely held belief that there are two methods, one popular, one exact. In a polemical work directed against Everard Digby, a prominent Cambridge Aristotelian and opponent of Ramism,[63] Temple claims: 'Those that attain no elegance of the discipline, and those that go to the academy use the... same precepts in searching for the truth and the same laws of inference in judging. In this way, method is not restricted to certain persons, or attached to certain things...'[64] For this reason, Temple also ridicules the Aristotelian distinction between 'being more known for us' and 'being more known according to nature': 'What if a practitioner of medicine, when categorizing beneficial things, would say that some of them were more beneficial for human beings, others however more beneficial

[61] On English Ramism and its critics, see Oldrini, 'Les Stratégies du combat chez Ramus et les Ramistes', pp. 34–40; Daniel, 'The Ramist context of Berkeley's philosophy'; Feingold, 'English Ramism: A Reinterpretation'; Boran, 'Ramism in Trinity College, Dublin in the Early Seventeenth Century'.

[62] Temple, *P. Rami Dialecticae libri duo, Scholiis Guilielmi Tempelli Cantabrigiensi illustrati*, pp. 1–2: '*E cuius... rei observatione Logicae artis praecepta fluxerunt? Certe ex observatione naturalis rationis, comprehensae quidem & expressae cum aliorum in republica praestantium hominum dictis & scriptis, tum poetarum, oratorum, philosophorum, theologorum sententiis ac commentationibus. Hi enim omnes naturali & divinitus tributa ratione usi sunt, id est, rei causas & effecta exposuerunt.*'

[63] See Digby, *De duplici methodo libri duo*.

[64] [William Temple,] *Admonitio Francisci Mildapetti Navarreni*, p. 55: '*Eisdem... praeceptis in veritate pervidenda, eisdem legibus in judicio consequentiae uti possunt & illi, qui nullius elegantiam disciplinae attigerunt, & illi qui Academiam profitentur. Sic methodus non est certis definita personis, non ad res certas alligata...*'

according to nature?... Or if a musician would claim that of two given sounds one is sharper for us, the other sharper according to nature...?'[65] According to Temple, in logic there is nothing that is 'more known according to nature' since all that can be known in logic derives from the description of the rules of natural reason.

When Clauberg, Arnauld, and Nicole object to the Ramist view of maxims that maxims cannot serve to increase our knowledge of particulars, they misidentify the target of their criticism. For Ramus and his followers, maxims, like other *loci*, are not knowledge-increasing tools. According to the Ramists, maxims, like other *loci*, are formulated by reflecting on what is going on in our practice of forming new arguments. Maxims describe only how human beings endowed with natural reason structure and increase their knowledge. A maxim such as 'Everything is what it is and not something else' describes the logical structure common to a proposition such as 'Red is not blue' and other propositions of the same logical form. Propositions that instantiate the logical structure described by the maxim are formed correctly; propositions that are contrary to the logical structure described by the maxim are formed incorrectly. According to Ramus and his followers, logical and metaphysical *loci* make explicit those rules that the mind already applies in everyday argumentative practice without reflecting on them. Reflecting on the actual argumentative practice of human beings—in particular, as exemplified by the great poets and orators—leads us to become aware of the role that these maxims play in our pre-reflective practice. Reflection thus leads to an awareness of normative components that are at work in our everyday reasoning. Giving an explicit formulation to maxims remains purely descriptive, because it does not introduce anything that was not implicitly presupposed in our everyday way of speaking and reasoning.

Even if Locke did not target the Ramists, his chapter on maxims misses the complexity of the field of existing alternative conceptions of the logical role of maxims. Because Locke's critique of maxims is closely similar to Cartesian criticisms of the Ramist tradition, what Locke objects against the Scholastic view is insufficient to exclude alternatives to the Scholastic view other than his own. Ramus does not think that our argumentative practices should be derived from reflecting on maxims and other rules of reasoning. According to his view, maxims do not fulfil any of the roles that Locke thought that 'Scholastic men' would ascribe to them: Maxims are not meant to increase the probability of our particular knowledge claims; they do not make anything certain that was not certain before; and they do not function as starting points for deductive arguments. Rather, according to Ramus maxims and other rules of reasoning describe the structure of the argumentative practices that we have already mastered. Ramus and the Ramists have a clear and interesting answer to the question of what the epistemic benefit of expressing maxims might be: Maxims describe formal aspects of actual examples of reasoning. But not any examples of reasoning will do for this purpose. This is why, according to Ramus and the Ramists, we have to analyse examples

[65] Ibid., p. 64.

of reasoning in outstanding authors. Due to the paradigmatic function of such examples, expressing maxims helps to stabilize our own argumentative practice: Forms of argument that correspond to the exemplary forms of argument are permissible, forms of argument that do not correspond to the exemplary forms are not permissible. Due to their connection with exemplary cases of reasoning, maxims at the same time function as descriptions of actual cases of reasoning and as criteria of correct reasoning.

Locke does not address the Ramist tradition in any explicit way. One reason for his passing over Ramism in silence may have been that he shared some of the methodological assumptions implicit in the Cartesian tradition. Methodological rules, ever since the publication of the *Regulae*, were meant to direct the mind towards novel discoveries. The compilation of methodological rules that Arnauld and Nicole give in the fourth book of the *La Logique* is characterized as furthering such a goal,[66] and Locke implicitly subscribes to a similar view of the function of methodological rules when he remarks that Newton reached all of the insights in his *Principia* without reflecting once upon maxims.[67] Tautologically, if the task of philosophy is defined as leading to novel discoveries, maxims do not contribute to what philosophy should do. But applying the Cartesian view of the function of methodological rules does not by itself constitute an argument that shows what is wrong with the analytic, descriptive, and criterial conception of maxims defended by the Ramists. Maybe Locke's criticism of the Scholastic, proof-theoretic view of maxims is as cogent as possible. And maybe he has accurately diagnosed the didactic and polemical functions that maxims can fulfil. Nevertheless, since he does not take the Ramist alternative into account, the argument leading from his critique of the Scholastic view of maxims to the view that the only functions that maxims can fulfil are expository and polemic appears to be less than conclusive.

4. Conclusion

The foregoing considerations have placed Locke's critique of maxims within the context of Cartesian criticisms of maxims. Locke did not necessarily derive his critique of maxims from the Port-Royal *Logic*, and Arnauld and Nicole probably did not derive his own critique of maxims from Clauberg. However, broadening the context of Locke's remarks on maxims makes it clear that Locke shares an overall argumentative strategy with the Cartesians: Like Clauberg, Arnauld, and Nicole, he argues that maxims are unsuitable for increasing human knowledge and hence can play a role only for the purposes of teaching and ending controversies. Such a line of argument may be effective when, as in the case of Locke, it is directed against the Scholastic conception of maxims. The same line of argument, however, is ineffective when, as in Cartesian logic, it is directed against the Ramist conception of maxims. Ramus shares with his critics the view that maxims cannot increase knowledge; but does not share the view that maxims, for this reason, are relevant only for didactic and polemical purposes. Rather, he holds that

[66] Arnauld and Nicole, *La Logique*, IV, 7.

[67] E IV. vii. 3, p. 599.

maxims, like other dialectical topics, describe rules that we apply in reasoning even before reflecting on these rules. Strangely enough, this point was missed by Clauberg, Arnauld, Nicole, and Locke alike. It is, I suggest, a metaphilosophical point: In his discussion of maxims, Ramus indicates a sense in which dialectics (as the part of philosophy concerned with maxims) can be understood as a descriptive discipline—as a discipline that is concerned with the analysis of (model cases of) everyday reasoning. Such a conception of the task of dialectics is left intact by the insight that maxims cannot be useful in the context of the logic of invention when 'invention' is understood in the sense of an expansion of knowledge. And this is so because Ramist 'invention' does not aim at creating new knowledge but rather at making us aware of what we implicitly have always known.

References

Arnauld, Antoine, and Nicole, Pierre. *La Logique ou L'Art de penser*. Édition critique par Pierre Clair et François Girbal. Paris: Presses Universitaires de France, 1965.

Arnauld, Antoine, and Nicole, Pierre. *La Logique, ou l'art de penser*. Édition critique par Dominique Descotes. Paris: Honoré Champion, 2011.

Ayers, Michael. *Locke. Volume I: Epistemology*. London and New York: Routledge, 1993.

Bacon, Francis. 'De dignitate et augmentis scientiarum', in James Spedding, Robert Leslie Ellis, and Douglas Denon Heath, eds., *The Works of Francis Bacon*. London: Longman, 1861: vol. 1.

Bacon, Francis. 'Maxims of the Law', in James Spedding, Robert Leslie Ellis, and Douglas Denon Heath, eds., *The Works of Francis Bacon*. London: Longman, 1861: vol. 7, pp. 307–87.

Bacon, Francis. *The Advancement of Learning*, ed. Michael Kiernan. Oxford: Clarendon Press, 2003.

Bacon, Francis. 'Novum organum', in Francis Bacon, *Novum organum and Associated Texts*, trans. Graham Rees. Oxford: Clarendon Press, 2004.

Beurhaus, Friedrich. *Paedagogiae logicae... pars secunda, editio locupletior*. Cologne: Maternus Cholinus, 1588.

Blank, Andreas. 'Ramus and Leibniz on Analysis', in *Leibniz: What Kind of Rationalist?*, ed. Marcelo Dascal. Dordrecht: Springer, 2008: 155–66.

Bohatec, Josef. *Die cartesianische Scholastik in der Philosophie und reformierten Dogmatik des 17. Jahrhunderts*, 1st edn. Leipzig, 1912; Hildesheim: Olms, 1966.

Boran, Elizabethanne. 'Ramism in Trinity College, Dublin in the Early Seventeenth Century', in Mordechai Feingold, Joseph S. Freedman, and Wolfgang Rother, eds., *The Influence of Petrus Ramus: Studies in Sixteenth and Seventeenth Century Philosophy and Sciences*... Basel: Schwabe, 2001: 177–99.

Clauberg, Johannes. *Defensio cartesiana*. Amsterdam: Elzevier, 1652.

Clauberg, Johannes. *Initiatio philosophi sive dubitatio cartesiana: Ad Metaphysicam certitudinem viam aperiens*. Lugdunum Batavorum and Duisburg: Adrian Wyngaerden, 1655.

Clauberg, Johannes. *Logica vetus et nova*, 3rd edn. [Nuremberg:] Schuerer; Sulzbach: Lichtenthaler, 1685.

Daniel, Stephen. 'The Ramist Context of Berkeley's Philosophy', *British Journal of the History of Philosophy*, 9 (2001), 387–412.

Dear, Peter. *Mersenne and the Learning of the Schools*. Ithaca, NY: Cornell University Press, 1988.

Digby, Everard. *De duplici methodo libri duo, unicam P. Rami methodum refutantes*. Frankfurt: Johannes Wechel, 1589.

Feingold, Mordechai. 'English Ramism: A Reinterpretation', in Mordechai Feingold, Joseph S. Freedman, and Wolfgang Rother, eds., *The Influence of Petrus Ramus: Studies in Sixteenth and Seventeenth Century Philosophy and Sciences*. Basel: Schwabe, 2001: 127–76.

Goclenius, Rudolph. *Lexicon philosophicum*. Marburg: Egenolph, 1613.

Harrison, John, and Laslett, Peter. *The Library of John Locke*, 2nd edn. Oxford: Clarendon Press, 1971.

Lagrée, Jacqueline. 'Sens et vérité chez Clauberg et Spinoza', *Philosophiques*, 29 (2002), 121–38.

Lentulus, Cyriacus. *Nova Renati Descartes sapientia: faciliori quam antehac methodo detecta*. Herborn: [no publisher], 1651.

Locke, John. *An Essay Concerning Human Understanding*, edited with an introduction, critical apparatus, and glossary by Peter H. Nidditch. Oxford: Clarendon Press, 1975.

Locke, John. *Drafts for the Essay Concerning Human Understanding, and Other Philosophical Writings*, ed. Peter H. Nidditch and G. A. J. Rogers. Vol. 1: Drafts A and B. Oxford: Clarendon Press, 1990.

Locke, John. 'Maxims.' MS Locke e.1, Oxford, Bodleian Library. The Locke Digital Project <http://www.digitallockeproject.nl>.

McKenna, Anthony. 'La Composition de la Logique de Port-Royal', *Révue philosophique de la France et de l'étranger*, 176 (1986), 183–206.

Mersenne, Marin. *La Verité des sciences. Contre les septiques ou Pyrrhoniens*. Paris: Toussainct du Bray, 1625.

Oldrini, Guido. 'Les Stratégies du combat chez Ramus et les Ramistes', in Kees Meerhoff and Jean-Claude Moisan, eds., with the collaboration of Michel Magnien, *Autour de Ramus: Le Combat*. Paris: Honoré Champion, 2005: 15–45.

Ramus, Petrus. *Institutionum dialecticarum libri tres... Audomari Talei praelectionibus illustratae*. Basileae: Nicolaus Episcopus Iunior, 1545.

Revius, Jacobus. *A Theological Examination of Cartesian Philosophy: Early Criticisms* (1647), ed. Aza Goudriaan. Leiden and Boston: Brill, 2002.

Robinet, André. 'La Postérité cartésienne du combat raméen et ramiste. Le lieu de lieu: lieu, étendue, espace', in Kees Meerhoff and Jean-Claude Moisan, eds., with the collaboration of Michel Magnien, *Autour de Ramus: Le Combat*. Paris: Honoré Champion, 2005: 107–35.

Roy, Bernard. 'Reasoned Grammar, Logic, and Rhetoric at Port-Royal', *Philosophy and Rhetoric*, 32 (1999), 131–45.

Savini, Massimiliano. 'L'Insertion du cartésianisme en logique: la *Logica vetus & nova* de Johannes Clauberg', *Révue de métaphysique et de morale*, 49 (2006), 73–88.

Savini, Massimiliano. *Johannes Clauberg:* Methodus Cartesiana *et ontologie*. Paris: Vrin, 2011.

Schuurman, Paul. 'General Introduction', in John Locke, *Of the Conduct of the Understanding*, edited with a general introduction, historical and philosophical notes, and critical apparatus by Paul Schuurman. Thesis (Ph.D.), University of Keele, 2000, pp. 15–126.

Schuurman, Paul. *Ideas, Mental Faculties, and Method: The Logic of Ideas of Descartes and Locke and its Reception in the Dutch Republic, 1630–1750*. Leiden: Brill, 2004.

[Temple, William.] *Admonitio Francisci Mildapetti Navarreni, ad Everardum Dygbeium Anglum, De unica P. Rami methodo, rejectis ceteris, retinenda*. Frankfurt: Johannes Wechel, 1589.

Temple, William. P. *Rami Dialecticae libri duo, Scholiis Guilielmi Tempelli Cantabrigiensi illustrate*, 3rd edn. Frankfurt: Petrus Fischer, 1605.

Trevisani, Francesco. *Descartes in Germania. La ricezione del cartesianesimo nella Facolta filosofica e medica di Duisburg (1652–1703)*. Milan: FrancoAngeli, 1992.

Trevisani, Francesco. 'Johannes Clauberg e l'Aristotele riformato', in Guido Canziani and Yves Charles Zarka, eds., *L'interpretazione nei secoli XVI e XVII. Atti del Convegno internazionale di studi Milano (18–20 novembre 1991), Parigi (6–8 dicembre 1991)*. Milan: FrancoAngeli, 1993: 103–26.

van Ruler, J. A. *The Crisis of Causality: Voetius and Descartes on God, Nature and Change*. Leiden, New York, and Cologne: Brill, 1995.

Verbeek, Theo. *Descartes and the Dutch: Early Reactions to Cartesian Philosophy, 1637–1650*. Carbondale, Ill.: Southern Illinois University Press, 1992.

Verbeek, Theo. 'Johannes Clauberg: A Bio-bibliographical Sketch', in Theo Verbeek, ed., *Johannes Clauberg (1622–1655) and Cartesian Philosophy in the Seventeenth Century*. Dordrecht, Boston, and London: Kluwer, 1999: 181–99.

Viola, E. 'Scolastica e cartesianesimo nel pensiero di J. Clauberg', *Rivista di filosofia Neo-Scolastica*, 67 (1975), 247–66.

Wood, Neal. *The Politics of Locke's Philosophy: A Social Study of 'An Essay concerning Human Understanding'*. Berkeley: University of California Press, 1983.

Yolton, John W. *The Two Intellectual Worlds of John Locke: Man, Person, and Spirits in the Essay*. Ithaca, NY, and London: Cornell University Press, 2004.

12

Locke and Malebranche

Intelligibility and Empiricism

Nicholas Jolley

Throughout his philosophical career Locke seems to have shown an intense dislike for what we might call 'divine direct assistance' models of knowledge; that is, he opposes all theories that explain the acquisition of ideas, beliefs, and principles in terms of God's role in implanting them in our minds. Perhaps the most obvious instance of Locke's hostility to such theories is the polemic against enthusiasm, which was added in a late edition of the *Essay* (IV. xix); here Locke criticizes those religious sectarians who claim that, as God's peculiar people, they are the recipients of immediate divine revelations. But Locke is also concerned to target more strictly philosophical adherents of the 'divine direct assistance' model. Thus in his polemic against nativism Locke repeatedly characterizes the theory as holding that God himself has inscribed certain ideas and principles on our minds. Even after publishing his own positive empiricist theory of idea-acquisition in the *Essay*, he returns to the issue towards the end of his career. In the 'Examination of P. Malebranche's Opinion of Seeing All Things in God' Locke mounts a detailed critique of the theory that God directly displays his own ideas to us, and that it is through such divine illumination[1] that we perceive bodies. In opposition to such philosophical theories Locke argues that we derive all our ideas from experience and acquire knowledge by the use of our natural faculties.

But why should we suppose that Locke's own empiricist account of ideas is the correct one, or even epistemically superior to its rivals? This question is one that has received relatively little attention in the literature. In this chapter I wish to address this issue in the light of Locke's examination of the Malebranchian theory of vision in God. In particular, I wish to raise the issue of whether some of Locke's criticisms of Malebranche's doctrine do not tell equally against Locke's own empiricist theory.

[1] As Martine Pécharman has pointed out to me, the charge that Locke's and Descartes's theory of ideas led to enthusiasm was made by John Sergeant in his *Solid Philosophy Asserted*, Reflection Nineteenth, para. 12, 377–9.

In one way such a line of criticism is not unfamiliar. Scholars have sometimes noted that in this work Locke advances one of the classic objections to the Representative Theory of Perception.[2] Thus it has been asked how Locke could have defended a doctrine whose weaknesses he himself was able to expose so clearly.[3] Some critics have risen to Locke's defence on this score by pointing to key differences between Malebranche's version of the Representative Theory of Perception and the version to which Locke himself is supposedly committed; thus it has been argued that Locke's theory is a causal one which, unlike the Malebranchian theory, gives bodies a key role to play in the production of ideas.[4] In a similar vein I propose to ask whether Locke's empiricist theory of ideas has the resources to counter the criticisms that he himself advances of the doctrine of vision in God. Although I pay some attention to the merits of Locke's critique, I am more concerned with the issue of what light it throws on the epistemic status of his own theory of ideas.

Some of the key themes of Locke's 'Examination of Malebranche's Opinion of Seeing All Things in God' emerge in the exchange of letters with William Molyneux. In a letter encouraging Locke to proceed with this project Molyneux writes:

> I look upon Malbranches Notions, or rather Platos, [as] in this particular perfectly unintelligible; And if you will Ingage in a Philosophick Controversy, you cannot do it with more advantage than in this matter. (Molyneux to Locke, 18 April 1693, LC 4, p. 668, letter 1622)

Locke reports back on the critique he has written in a way that echoes and develops Molyneux's position:

> I have also examined Pere Malbranche's opinion concerning seeing all things in God, and to my own satisfaction laid open the vanity, and inconsistency, and unintelligibleness of that way of explaining humane understanding.
>
> (Locke to Molyneux, 8 March 1695, LC 5, p. 287, letter 1857)

It might be tempting to suppose that Locke is simply engaging in vague abuse here, but that, I think, would be a mistake; by alluding to the 'vanity', 'inconsistency', and 'unintelligibleness' of the theory Locke has some rather specific criticisms in mind in this passage. Thus Locke holds that the doctrine of vision in God is unintelligible by virtue of the fact that it purports to explain our having ideas in terms of a union between the human mind and God. According to Locke, such a theory employs the concept of union beyond its legitimate sphere of application; the concept of union applies to

[2] On this issue see Woozley (ed.), *John Locke: An Essay Concerning Human Understanding*, introduction. I am grateful to Henry Allison, Jan-Erik Jones, and two anonymous referees for their criticisms of earlier drafts of this chapter; I should also like to thank Philippe Hamou, Martine Pécharman, and the participants of the 2012 Lille conference on Locke and Cartesianism for their helpful comments.

[3] See, for instance, LW 9, p. 221: '[How] can he know that there is any such real being in the world as the sun? Did he ever see the sun? No, but on occasion of the presence of the sun to his eyes, he has seen the idea of the sun in God, which God has exhibited to him; but the sun, because it cannot be united to the soul, he cannot see. How then does he know that there is a sun which he never saw?'

[4] See Matthews, 'Locke, Malebranche, and the Representative Theory of Perception'.

bodies but it has no application to the mind and God (LW 9, p. 213). Locke holds that Malebranche is guilty of inconsistency inasmuch as he claims that the union of mind and God is itself sufficient to explain the perception of ideas, but he also explicitly denies this when he further says that God must discover his ideas to us (LW 9, p. 228); thus God has the idea of the human mind in himself, but we have no epistemic access to such an idea in contrast to the ideas of bodies or intelligible extension. Locke further holds that the doctrine of vision in God is vain in the sense that it is explanatorily empty. For to augment the theory of the union of mind and God by saying that God must discover ideas to us is to say nothing more than Locke and everyone else is prepared to concede; it conveys no sense of the 'mechanism' by means of which God acts on the mind in the perception of ideas (LW 9, pp. 228–9).

Perhaps the most important criticism which Locke does not cite in his report to Molyneux concerns Malebranche's overall strategy of argument for the vision in God. Here we should note that Malebranche's strategy is not constant over time. In late works such as the *Dialogues on Metaphysics and on Religion* Malebranche argues for his theory from the properties of ideas: he first establishes that we perceive bodies not directly but by means of ideas that represent them, and then argues that all ideas have properties—for example, immutability, eternity, necessity, and infinity—that are such that God alone can be their locus (DM 2, p. 1, OCM 12, pp. 50–1; JS 20–1). In *The Search After Truth*, the target of Locke's critique, although he similarly begins by establishing the Representative Theory of Perception, in other respects Malebranche adopts a very different approach: he employs the strategy of an argument by elimination. Malebranche introduces this argument by means of a '*classification of all the ways external objects can be seen*':

> We assert the absolute necessity, then, of the following: either (a) the ideas we have of bodies and of all other objects we do not perceive by themselves come from these bodies or objects; or (b) our soul has the power of producing these ideas; or (c) God has produced them in us while creating the soul or produces them every time we think about a given object; or (d) the soul has in itself all the perfections it sees in bodies; or else (e) the soul is joined to a completely perfect being that contains all intelligible perfections, or all the ideas of created beings.
>
> (SAT 3. 2. 1, OCM 1, p. 417: LO 219)[5]

In subsequent chapters Malebranche seeks to refute the first four theories so that (e), the vision in God, alone remains in the field. The exact nature of the theories that are the targets of Malebranche's arguments is controversial, but for our purposes the issue can be set on one side; what is important is that Malebranche offers an indirect proof of his theory of vision in God.[6]

[5] The letters in brackets are introduced by Lennon and Olscamp. They are retained here for the reader's convenience.

[6] For an illuminating discussion of the targets of Malebranche's arguments, see Nadler, *Malebranche and Ideas*, pp. 108–37.

In his 'Examination' Locke complains that Malebranche's proof is defective. It is simply an *argumentum ad ignorantiam*, that is, an argument from our ignorance of any alternative ways in which we might perceive bodies:

> Which argument, so far being only 'argumentum ad ignorantiam,' loses all its force as soon as we consider the weakness of our minds, and the narrowness of our capacities, and have but humility enough to allow, that there may be many things which we cannot fully comprehend, and that God is not bound in all he does to subject his ways of operation to the scrutiny of our thoughts, and confine himself to do nothing but what we must comprehend. (LW 9, p. 212)

Now Locke himself is right that the soundness of an argument by elimination depends on an exhaustive enumeration of the alternative options. Although some commentators, such as Steven Nadler, have tried to come to his assistance, it cannot be said that Malebranche himself provides a logical basis for the division; thus he does not succeed in demonstrating that the enumeration is complete.[7] But Locke's own critique of Malebranche on this topic seems unsatisfactory as it stands. First, as Charles McCracken points out, Locke cannot legitimately appeal to the weakness of our minds and the narrowness of our capacities to rule out all disjunctive arguments; consider, for example, the case of a disjunctive syllogism where the disjuncts are contradictories.[8] Secondly, Locke's argument against Malebranche seems to be in some danger of begging the question. For how would Locke justify the premise that our minds are weak and our capacities are narrow? Locke might well try to do so by citing the fact that key metaphysical ideas—such as the ideas of substance, identity, and infinity—are not clear and distinct, as the Cartesians suppose, but obscure and confused; if pressed, he would in turn justify these claims by saying that the content of our ideas is limited by their humble origins in the data of sensation and reflection. But if, in his argument against Malebranche, Locke is forced to invoke his own empiricist theory to justify the premise that our faculties are limited, then he is guilty of a *petitio principii*.

Locke himself seems to be aware of the danger of begging the question against one's opponent when the quality of the human understanding is at issue. For Locke criticizes Malebranche for his confident assertion that his enumeration is complete by saying that it must be built on the good opinion of our capacities that God cannot make the creatures operate but in ways conceivable to us (LW 9, p. 214). To ascribe such a good opinion of our faculties to Malebranche may seem strange to readers familiar with his later works where he goes out of his way to emphasize that our minds and their modalities are but darkness unless assisted by the light of divine illumination (DM 2, p. 7, OCM 12, p. 67; JS 35). But it is arguably not so strange or inaccurate as an account of Malebranche's position in *The Search After Truth* where he seems to hold that the mind is endowed with a Cartesian faculty of pure intellect; such a faculty may not be capable

[7] Nadler, *Malebranche and Ideas*, p. 139.
[8] McCracken, *Malebranche and British Philosophy*, p. 127.

of drawing ideas out of its own depths, but it is capable of apprehending ideas in God.[9] But whatever the accuracy of his reading of Malebranche, Locke's point seems clear: Malebranche himself has no non-question-begging way of justifying his assumption that our faculties are good enough to allow us to give an exhaustive enumeration of the ways in which we might perceive bodies.

Locke's opposition to Malebranche's strategy of an argument by elimination is instructive, for it might seem on the evidence of the 'Examination' that Locke has the resources for such an argument for his own positive empiricist theory. Thus Locke has decisively ruled out both strong and weak versions of the nativist hypothesis in the *Essay*: the strong version, which asserts the actual possession of innate ideas, is empirically false, and the weak version, which asserts a capacity for such possession, is trivially true. In the 'Examination of Malebranche' he not only rules out the vision in God but also endorses at least some of Malebranche's criticisms of the Scholastic theory of species; indeed, in one place he is prepared to write of the 'learned gibberish of the Schools' (LW 9, p. 215). We can imagine a philosopher—call him Locke*—who thus employs an argument by elimination in order to establish the empiricist hypothesis; such a philosopher, let us say, is without Locke's own scruples about the weakness of the human understanding. In contrast to Locke*, Locke cannot avail himself of an argument by elimination to prove the empiricist theory of ideas, and indeed, he may not think that the theory is strictly capable of proof, whether direct or indirect. Nonetheless, it is natural to ask what reasons he has for thinking that the empiricist theory is at least epistemically superior to its rivals.

Since Locke criticizes the vision in God for its emptiness, inconsistency, and unintelligibility, he must surely think that the empiricist theory of ideas is free from these defects: in contrast to the vision in God, it offers an explanation of our ideas that is not only consistent and intelligible but also non-circular. Whether Locke's theory in fact avoids circularity and inconsistency is a large and controversial issue that I have not space to discuss here. In what follows I propose to focus on the issue of whether his theory is more intelligible than its rivals, and by what criteria.

A promising way of approaching this issue is by means of the corpuscularian hypothesis. There are two reasons why such an approach seems promising. In the first place, it is clear that the empiricist theory of ideas is to some extent modelled on this hypothesis, and is supposed to gain credit from its success. Both theories, for instance, posit the existence of simples and explain composite entities as arising out of the simples in various ways. Some evidence for the claim that the empiricist theory is modelled on the corpuscularian hypothesis is furnished by the use of the alphabet analogy. In *About The Excellency and Grounds of the Mechanical Hypothesis* Boyle had employed this analogy in order to defend the hypothesis against the objection that it was too meagre in its resources to explain all the variety of physical phenomena that it was required to

[9] On this issue see Jolley, 'Intellect and Illumination in Malebranche'.

explain; Boyle responds by drawing attention to the fact that the mere twenty-four (*sic*) letters of the alphabet are sufficient to produce all the great works of literature in the world.[10] In the same way Locke invokes the alphabet analogy in order to defend the empiricist theory against a parallel objection:

> Nor will it be so strange, to think these few simple *Ideas* sufficient to employ the quickest Thought, or largest Capacity; and to furnish the Materials of all that various Knowledge, and more various Fancies and Opinions of all Mankind, if we consider how many Words may be made out of the various composition of 24 letters; or if going one step farther, we will but reflect on the variety of combinations may be made, with barely one of the above-mentioned *Ideas, viz.* Number, whose stock is inexhaustible, and truly infinite: And what a large and immense field, doth Extension alone afford the Mathematicians? (E II. vii. 10)

Secondly, the virtue of the corpuscularian hypothesis that Locke explicitly advertises is precisely its superior intelligibility compared with its rivals such as the Scholastic theories. According to a well-known passage in IV. iii. 16 of the *Essay*, the corpuscularian hypothesis, while it fails to achieve the status of knowledge or *scientia*, is that which 'is thought to go furthest in an intelligible Explication of the Qualities of Bodies'. Now as we might expect, Locke's appeal to the intelligibility of the corpuscularian hypothesis is not original with him; it has a precedent in the work of his friend and mentor Boyle. But, as Edwin McCann remarks, Boyle, 'for all his boosterism, makes appeal to completely undefined and unexplained standards of intelligibility' in connection with the corpuscularian hypothesis.[11] And Locke himself is not much more forthcoming about what the criteria of intelligibility are supposed to be.

One plausible candidate as a criterion is what we may call, taking our cue from Peter Anstey, the Familiarity Condition;[12] that is, the corpuscularian hypothesis attributes no determinable properties to the unobservable simple entities (the corpuscles) with which we are not already familiar at the macroscopic level of compound bodies such as tables and chairs. Thus the corpuscles, along with tables and chairs, are endowed with such properties or qualities as bulk (size), figure (shape), solidity, motion or rest, and position. In this respect the corpuscularian hypothesis is superior to Scholastic-style explanations which appeal to occult qualities and substantial forms that we do not encounter at the level of macroscopic phenomena.

When we turn to Locke's theory of ideas, it is clear that there is no straightforward analogue of the Familiarity Condition in physics. As we have seen, in connection with the corpuscularian hypothesis, the Familiarity Condition may be formulated in terms of a claim about unobservable entities; it is the corpuscles that are too

[10] *About The Excellency and Grounds of the Mechanical Hypothesis*, p. 142.

[11] McCann, 'Locke's Philosophy of Body', p. 59.

[12] Anstey, *John Locke and Natural Philosophy*, pp. 157ff. Anstey states the Familiarity Condition as follows: 'all explanations of the unobserved must be made in terms of properties and causes with which we are familiar'. I find this formulation puzzling, since what is at issue is surely rather the explanation of observable, macroscopic phenomena in terms of unobservable bodies or corpuscles that have only determinable properties with which we are familiar.

small to be observed with the naked eye or even with the use of microscopes that have no determinable properties which are not familiar at the level of macroscopic bodies. By contrast, the empiricist theory of ideas of course is not supposed to traffic in unobservable entities: the simples are supposed to be the data of conscious experience. It is true that in order to defend Locke against criticisms of his theory of simple ideas some commentators have been tempted to resort to a distinction between absolute and relative simplicity.[13] One might say that the simples that are the data of conscious experience are only relatively simple, thus leaving open the possibility that there are absolute simples that are not given in conscious experience. But any such resort to a doctrine of unobservable absolute simples runs into trouble from Locke's commitment to the Cartesian thesis that the mind is transparent to itself: there is nothing in the mind of which we are not conscious (E I. ii. 19; cf. Descartes, AT 3, p. 273). Ironically, the Familiarity Condition when stated as a claim about unobservables would have more purchase in the case of a theory of perception such as Leibniz's; Leibniz postulates the existence of *petites* or unconscious perceptions, and he explicitly invokes the analogy with the insensible corpuscles of physics (NE, Preface, A 6, p. 6; RB 56). Such insensible perceptions could satisfy the Familiarity Condition by virtue of having representational content and intensive magnitude in common with conscious perceptions or ideas. But as we have seen, this way of satisfying the Familiarity Condition on intelligibility is not available to Locke, who is committed to the Cartesian thesis of the transparency of the mental, at least with regard to occurrent items.

We can make progress with the issue if we suppose that, with respect to the theory of ideas, Locke is committed to a Familiarity Condition of the following form:

All mental content that is not given in experience (including general ideas) is to be explained in terms of mental content and operations with which we are familiar.

Now it is surely plausible to suppose that in Locke's eyes Descartes's theory of innate ideas violates this condition. Consider, for instance, the classic statement of his theory in the *Comments on a Certain Broadsheet*:

I have never written or taken the view that the mind requires innate ideas which are something distinct from its own faculty of thinking. I did, however, observe that there were certain thoughts within me which neither came to me from external objects nor were determined by my will, but which came solely from the power of thinking within me; so I applied the term 'innate' to the ideas or notions which are the forms of these thoughts in order to distinguish them from others, which I called 'adventitious' or 'made up'. This is the same sense as that in which we say that generosity is 'innate' in certain families, or that certain diseases such as gout or stones are innate in others: it is not so much that the babies of such families suffer from these diseases in their mother's womb, but simply that they are born with a certain 'faculty' or 'tendency' to contract them. (AT 8B, p. 357; CSM 1, pp. 303–4)

[13] See Stewart, 'Locke's Mental Atomism and the Classification of Ideas (1)'. Cf. Jolley, *Locke: His Philosophical Thought*, p. 46.

According to Descartes, then, we have purely intellectual, occurrent thoughts that are proximately caused by an innate mental disposition. And for Locke such a theory is suspect on two grounds. First, according to Locke, it is simply false to say that we have purely intellectual thoughts of God and triangles, for instance; experience turns up no ideas which do not have, as he picturesquely calls it, a 'tangue of the Cask' (E II. i. 17). In Locke's eyes the commitment to innate mental dispositions may also violate the Familiarity Condition. Locke cannot of course say that we have no experience of dispositional properties, for he is committed to holding that we have ideas of secondary qualities, and secondary qualities are dispositional properties of a certain kind; they are powers in bodies to produce ideas in us (E II. viii. 10). But Locke can arguably drive a wedge between the two cases by saying that secondary qualities are powers that are grounded in the categorical properties of bodies; by contrast, the mental dispositions of which Descartes writes are not grounded in any categorical properties, mental or physical.

Whether Cartesian nativism in fact violates the Familiarity Condition may be disputed; for it is open to Descartes to say that we do in fact experience purely intellectual thoughts of God and triangles, at least when we are thinking clearly and distinctly. There is thus a danger that the debate between Locke and Descartes on this issue will be simply a stalemate. But whatever the merits of Locke's critique, his hostility to this doctrine places us in a position to see how his empiricist theory of ideas may be supposed to satisfy the Familiarity Condition on intelligibility. Consider, for instance, the theory with which Locke seeks to replace Cartesian nativism. As I have argued elsewhere, Locke holds that, among its other advantages, his own theory of abstract ideas can do duty for Descartes's doctrine of innate ideas;[14] in particular, it can give an account of how ideas can be general without resorting to allegedly mysterious entities. For Locke, abstract ideas are grounded in the simple ideas that are given in experience and the complex ideas which, if not immediately given, are constructed out of the simples by a familiar mental process of combination. Indeed, according to some commentators such as Vere Chappell, at least some complex ideas are actually given in experience.[15] It is this criterion of intelligibility that Locke's theory of abstract ideas can be said to satisfy.

It is a slightly complicating fact that in different places in the *Essay* Locke seems to offer two rival accounts of the mental process by which abstract ideas are formed. In II. xi. 9 Locke offers what might be called a 'selective attention' account; on this view I have a particular experience of white, and I attend to that image only in respect of that feature in terms of which resembles my ideas of chalk and milk. By contrast, in III. iii. 7 Locke seems to offer what may be called an 'elimination' account of how my abstract

[14] Jolley, 'Lockean Abstractionism versus Cartesian Nativism.'

[15] 'My mind is no less passive with respect to this complex idea [of the apple] than it is with respect to the simple ideas that compose it.' Chappell, 'Locke's Theory of Ideas', p. 37. For criticism of Chappell's interpretation see Jolley, *Locke*, pp. 46–8.

ideas are formed. In terms of the example he gives there, we experience ideas of a number of individuals such as John, Paul, and Mary, and we eliminate everything except that which is common to them all; what is left is the abstract idea of a human being. But on either account a case can be made for saying that abstract ideas satisfy the Familiarity Condition. The satisfaction of this condition is perhaps especially clear in the case of the 'selective attention' account. On this account, to form the general idea of a triangle it is not necessary to postulate a schematic, indeterminate mental image as the 'elimination' account seems to do; it is sufficient to attend to the idea of a particular triangle that we already have, and to consider the idea in terms of those properties that it shares with other triangles.

The analogy, then, with the Familiarity Condition in physics is not to be found in an account of unobservable simples that have no determinable properties with which we are not familiar in experience; it is to be found in the theory of abstract, general ideas that serves as a rather systematic replacement for Cartesian nativism. It is of course an irony of the story that Locke's theory of abstract ideas was considered far from intelligible by one major philosopher. In the Introduction to the *Principles of Human Knowledge* Berkeley famously attacked the theory on the grounds that it claims that we can have ideas that combine inconsistent features; here Berkeley took his cue from Locke's perhaps unguarded remark that the idea of a triangle is neither oblique, nor rectangle, neither equilateral, nor scalenon, but 'all and none of these at once' (E IV. vii. 9). But Berkeley's critique arguably depends on a misreading of Locke's text; indeed, at least on the 'selective attention' account Locke's own theory of abstract ideas is close to the account of generality that Berkeley himself endorses.

It seems possible, then, to specify a sense in which the empiricist theory of ideas is supposed to satisfy the Familiarity Condition on intelligibility, even if the analogy with the corpuscularian hypothesis is not as strong as we could wish. Nonetheless, to say this is not to say that Locke has fully succeeded in vindicating his theory against the charge that it is in some way unintelligible. In fact, when we look at other aspects of the empiricist theory, we see that it is plagued by the same sorts of difficulties as attend Locke's philosophy of matter.

To set up the issue, let us consider another way in which, according to Locke, the corpuscularian hypothesis and its commitment to mechanism might be thought to be intelligible. According to some statements, Locke holds that it is a positive feature of the new theory of matter that there are intelligible a priori conceptual connections between the properties of the corpuscles and the macroscopic properties of bodies: such macroscopic properties are supposed to flow with geometrical necessity from the properties of the tiny corpuscles. Such a view seems to inform Locke's optimistic discussion of how, on the corpuscularian hypothesis, 'it is easie to be understood' how the same water can feel cold to one hand and warm to the other (E II. viii. 21). Yet, as Margaret Wilson has argued in a classic but controversial article, there are many passages where Locke seems to qualify his commitment to the view that the physical world is intelligible in this sense.[16] Thus Locke argues that we cannot understand how secondary qualities flow from the properties of the insensible particles; the existence of such a connection between a certain corpuscular structure and the quality of redness, for instance, must be simply ascribed to God's good pleasure. Moreover, in the wake of the Newtonian theory of universal gravitation, Locke comes to realize that bodies interact in ways that are not countenanced by Boylean mechanism:

> The gravitation of matter towards matter, by ways inconceivable to me, is not only a demonstration that God can, if he pleases, put into bodies powers and ways of operation above what can be derived from our idea of body, or can be explained by what we know of matter, but also an unquestionable and every where visible instance, that he has done so. (LW 4, pp. 467–8)

Now attempts have been made to reconcile such admissions with Locke's commitment to the intelligibility of the physical world, but even its proponents admit that though all may be intelligible to God, not everything is intelligible to us,

When we return to the empiricist theory of ideas, we find that it reproduces the difficulties that Locke acknowledges in his philosophy of matter. Let us introduce the topic by noting that, with regard to sensory ideas, Locke's empiricist theory involves two distinct components. On the one hand, the theory involves what we might call, following Jonathan Lowe, a constructivist component;[17] as we have seen, complex ideas of sensation result from mental operations performed on the simple ideas. On the other hand, the theory involves what we might call an interactionist component: simple ideas of sensory experience result from the action of bodies on the sense-organs, the brain, and ultimately on the mind. These two elements of the theory are of course logically independent. Thus it would be possible to hold both that all complex ideas are constructed out of simples and that these simple ideas are directly produced in the mind by God. Conversely, it would be possible to hold that all complex ideas are given in experience while maintaining that all ideas are produced by the action of bodies on the mind. It is interesting to note that Hume is an example of a philosopher who endorses the constructivist component of the empiricist theory but does not commit himself to the interactionist component; Hume professes to be simply agnostic about the causes of simple impressions (T I. I. ii). Locke, by contrast, is an example of an empiricist who endorses both components.

It is with respect to the interactionist component of the empiricist theory that Locke recognizes problems analogous to those that he acknowledges in the case of his philosophy of matter. It is true that, as in his philosophy of matter, Locke sometimes sounds an optimistic note: on occasion he writes rather breezily as if there were no difficulty in understanding how human beings endowed with eyes could come to have sensory ideas of colours: 'I imagine any one will easily grant, That it would be impertinent to suppose, the *Ideas* of Colours innate in a Creature, to whom God hath given Sight, and a Power to receive them by the Eyes from external Objects' (E I. ii. 1).

[16] Wilson, 'Superadded Properties: The Limits of Mechanism in Locke'.

[17] Lowe, *Locke*, p. 34.

In such a passage no account is taken of a well-known argument of Descartes's in *Comments on a Certain Broadsheet*; there Descartes argues from the Causal Likeness Principle for the thesis that even sensory ideas are innate in the mind (AT 8B, p. 359; CSM 1, p. 304). But in his 'Examination of Malebranche' Locke is more candid about the conceptual difficulties:

> Impressions made on the retina by rays of light, I think I understand; and motions from thence continued to the brain may be conceived, and that these produce ideas in our minds, I am persuaded, but in a manner to me incomprehensible. This I can resolve only into the good pleasure of God, whose ways are past finding out. (LW 9, p. 217)
>
> And if I should say, that it is possible God has made our souls so, and so united them to our bodies, that upon certain motions made in our bodies by external objects, the soul should have such or such perceptions of ideas, though in a way inconceivable to us; this perhaps would appear as true and as instructive a proposition as what is so positively laid down. (LW 9, p. 215)

Thus in his philosophy of matter Locke recognizes that bodies gravitate towards each other in ways inconceivable to us, and in his empiricist theory of ideas he recognizes that a key part of the process is incomprehensible and inconceivable to us; like gravitation, it must be ascribed to God's good pleasure.

Here it is interesting to note the contrast with Malebranche's own teachings. It is a striking fact that the doctrine of the vision in God circumvents all the problems posed by the production of ideas by bodies. Now for Malebranche, at least in his mature philosophy, ideas, unlike sensations, are not the sort of entities that could have efficient causes, since they are logical concepts or third-realm entities. But even if, for Malebranche, it made sense to suppose that they had causes, it would not be coherent to claim, as Locke does, that they are genuinely produced in the mind in a way that is incomprehensible to us. For Malebranche, it is of course of the very essence of the causal relation to be intelligible; 'a true cause as I understand it is one such that the mind perceives a necessary connection between it and its effect' (SAT 6. 2. 3, OCM 2, p. 316; LO 450). But the key point for our purposes is not the difference between a rationalistic and a non-rationalistic conception of causality; it is rather that it is a prima facie strength of the Malebranchian theory that it postulates nothing incomprehensible about our perception of the physical world. Locke of course will reply that the theory is in fact unintelligible, but when he is being candid he must acknowledge that his own theory recognizes that our perception of the physical world involves something incomprehensible and inconceivable.

Since Locke recognizes such difficulties in his empiricist theory of ideas, it is natural to ask whether this theory is epistemically on a par with the vision in God. It would be a mistake to jump to this conclusion. When Locke complains that Malebranche's vision in God is unintelligible, he is making a semantic claim about statements. To say, for instance, as Malebranche does, 'God is the place of spirits' is absurd, if it is taken literally. If, however, the claim is taken metaphorically, then it 'signifies literally nothing' (LW 9, p. 223). By contrast, when Locke recognizes the existence of problematic elements in his own theory he is not engaged in making semantic claims of this sort. To say that bodies gravitate towards each other and produce ideas in the mind is to say, quite meaningfully, that there are processes that we do not understand; incomprehensibility and inconceivability are predicated of features of the world, not of statements. Whatever the problems posed by the empiricist theory or the theory of universal gravitation, they are not literally meaningless.

But to say this is not to say that Locke is out of the wood; his critique of Malebranche may still do damage to his own empiricist theory of ideas. Consider the issue of the epistemic status of the empiricist theory of ideas by Locke's own lights. It is uncontroversial, for one thing, that the theory of ideas is not a case of *scientia* or demonstrative knowledge; Locke is not prepared to advance so strong a claim even for the corpuscularian hypothesis, and it is clear that the epistemic status of the theory of ideas rises no higher than that of the physical hypothesis on which it seems to be modelled. Moreover, we have seen that Locke cannot claim that the empiricist theory is established by an indirect proof such as an argument by elimination. With the presumable exception of arguments whose disjunctive premises are contradictories, Locke holds that when we modestly consider the weakness of our own minds we will see that no argument by elimination can carry any force. At most, then, Locke is in a position to say that the theory is the best of the available alternatives where the criteria for evaluating such hypotheses are not fully spelt out. But this seems a strikingly weak claim on behalf of the theory. In view of his insistence on the need for epistemic humility, it seems that Locke is forced to admit that it is possible, for all we know, that there are other ways in which we might come to possess ideas. It is true of course that alternative hypotheses on this subject might similarly be obliged to admit the existence of incomprehensible processes, but there seems to be no a priori reason why this should be so; perhaps a theory which postulates nothing incomprehensible has just not occurred to us yet. In view of such considerations it appears that, with some modification, one of Locke's own statements can be used against him: 'It will very little help to cure my ignorance that this is the best of four or five hypotheses proposed which are all defective, if this too has in it what is inconsistent with itself or unintelligible to me' (LW 9, p. 212). If we replace 'unintelligible' with 'inconceivable' or 'incomprehensible', it seems that Locke has stated a powerful criticism of his own empiricist theory of ideas.

References

Anstey, Peter R. *John Locke and Natural Philosophy*. Oxford: Oxford University Press, 2011.

Boyle, Robert. *About the Excellency and Grounds of the Mechanical Hypothesis*, in M. A. Stewart, ed., *Selected Philosophical Papers of Robert Boyle*. Manchester: Manchester University Press, 1979.

Chappell, Vere. 'Locke's Theory of Ideas', in V. Chappell, ed., *The Cambridge Companion to Locke*. Cambridge: Cambridge University Press, 1994: 26–55.

Jolley, Nicholas. 'Intellect and Illumination in Malebranche', *Journal of the History of Philosophy*, 32 (1994), 209–24.

Jolley, Nicholas. *Locke: His Philosophical Thought*. Oxford: Oxford University Press, 1999.

Jolley, Nicholas. 'Lockean Abstractionism versus Cartesian Nativism', in P. Hoffman, D. Owen, and G. Yaffe, eds., *Contemporary Perspectives on Early Modern Philosophy: Essays in Honor of Vere Chappell*. Toronto: Broadview Press, 2008: 157–71.

Lowe, E. Jonathan. *Locke on Human Understanding*. Abingdon and New York: Routledge, 2005.

McCann, Edwin. 'Locke's Philosophy of Body', in V. Chappell, ed., *The Cambridge Companion to Locke*. Cambridge: Cambridge University Press, 1994: 56–88.

McCracken, Charles J. *Malebranche and British Philosophy*. Oxford: Clarendon Press, 1983.

Matthews, H. E. 'Locke, Malebranche, and the Representative Theory of Perception', in I. C. Tipton, ed., *Locke on Human Understanding*. Oxford: Oxford University Press, 1977: 55–61.

Nadler, Steven. *Malebranche and Ideas*. Oxford and New York: Oxford University Press, 1992.

Sergeant, John. *Solid Philosophy Asserted, Against the Fancies of the Ideists*. London, 1697; repr. New York and London: Garland, 1984.

Stewart, Michael A. 'Locke's Mental Atomism and the Classification of Ideas (1)', *Locke Newsletter*, 10 (1979), 53–82.

Wilson, Margaret D. 'Superadded Properties: The Limits of Mechanism in Locke', *American Philosophical Quarterly*, 16 (1979), 143–50; repr. in M. Wilson, *Ideas and Mechanism*. Princeton: Princeton University Press, 1999: 196–208.

Woozley, Anthony D., ed., *John Locke: An Essay Concerning Human Understanding*. London: Fontana, 1964.

General Index